Charles R. O'Melia

August, 1999

# Time, Love, Memory

# Time, Love, Memory

A GREAT BIOLOGIST AND HIS QUEST
FOR THE ORIGINS OF BEHAVIOR

## JONATHAN WEINER

*Alfred A. Knopf*

NEW YORK    1999

Copyright © 1999 by Jonathan Weiner
All rights reserved under International and Pan-American
Copyright Conventions. Published in the United States by
Alfred A. Knopf, Inc., New York, and simultaneously in Canada
by Random House of Canada Limited, Toronto.
Distributed by Random House, Inc., New York.
www.randomhouse.com

Knopf, Borzoi Books, and the colophon are
registered trademarks of Random House, Inc.

Library of Congress Cataloging-in-Publication Data
Weiner, Jonathan.
Time, love, memory : a great biologist and his quest for the
origins of behavior / by Jonathan Weiner. — 1st ed.
p.    cm.
Includes bibliographical references and index.
ISBN 0-679-44435-1 (alk. paper)
1. Behavior genetics.    2. Benzer, Seymour.    I. Title.
QH457.W43    1999
591.5—dc21        98-43128
CIP

Manufactured in the United States of America
First Edition

*For two good friends:*
*my brother, Eric,*
*and John Bonner*

*Little Fly,*
*Thy summer's play*
*My thoughtless hand*
*Has brush'd away.*

*Am not I*
*A fly like thee?*
*Or art not thou*
*A man like me?*

—WILLIAM BLAKE,
"The Fly,"
from *Songs of Experience*

# Contents

# Time, Love, Memory

# PART ONE

# Occam's Castle

*Would it be too bold to imagine, that in the great length of time, since the earth began to exist, perhaps millions of ages before the commencement of the history of mankind, would it be too bold to imagine, that all warm-blooded animals have arisen from one living filament . . . possessing the faculty of continuing to improve by its own inherent activity, and of delivering down those improvements by generation to its posterity, world without end!*

—ERASMUS DARWIN,
*Zoonomia,* 1794

*My handwriting same as Grandfather.*

—CHARLES DARWIN,
a scribble in the M notebook, 1838

# CHAPTER ONE

---

# From So Simple a Beginning

*The ancient precept, "Know thyself," and the modern precept, "Study nature," become at last one maxim.*

—RALPH WALDO EMERSON,
"The American Scholar"

*The nearest gnat is an explanation.*

—WALT WHITMAN,
"Song of Myself"

SEYMOUR BENZER'S laboratory runs along two corridors of Church Hall at the California Institute of Technology in Pasadena. His private workroom is at the corner where the corridors meet. Here he keeps his own tools and trophies, and an owl's hours. It is a windowless room lined with plastic bins that Benzer labeled decades ago in his spidery black script: *Lenses, Mirrors, Needles, Wires, Pencils, Switches, Toothpicks, Pipe Cleaners*, anything and everything he might need for an experiment in the middle of the night, including *Teeth (Human and Shark)*.

The old gray benchtop is all test tubes and bottles: mostly standard-issue laboratory stock, but here and there a half-pint milk bottle with heavy scratched glass and antique advertising ("5 cents—Just a Little Better") stoppered with a foam-rubber cork. These tubes and bottles hold a sampling of the hundreds of mutants that Benzer and his students, his students' students, and his competitors have engineered.

The mutants are fruit flies, and their mutations have changed their behavior. One of them is *timeless*, a clock mutant. In a windowless room like this one, the fly seems to wake and sleep at random intervals, as if it has broken its covenant with day and night, so that day and night will not come at their appointed time. Another is *dissatisfaction,* a female mutant that does not like males and keeps flicking them away with her wings. Then there is *pirouette,* which moves at first in big arcs, then in smaller and smaller arcs, like certain problems in science, turning at last on a single point, until it sometimes starves to death.

In the seventeenth century, the French philosopher Blaise Pascal looked up at the night sky and then looked down at a mite, picturing "legs with joints, veins in its legs, blood in the veins, humors in the blood, drops in the humors, vapors in the drops," and onward and downward to the atoms. "The eternal silence of these infinite spaces fills me with dread," he wrote. He meant two infinite spaces, which he called the two infinites of science, one above and around him, the other below and inside him. Of the two infinites, the space that frightened him more was the space that he could not begin to see, the stardust of atoms that made up his very thoughts and fears and moved the fingers around his pen. "Anyone who considers himself in this way will be terrified at himself."

The twentieth century was a long spiral inward on Pascal's path, beginning with a single mutant fly in a milk bottle in the century's first years, and reaching the atoms that Pascal dreaded to see near the century's close. If the spiral leads where it now promises or threatens to lead, this may be remembered as one of the most significant series of discoveries since science began, matching the discoveries of twentieth-century physics. In the universe above and around us, physics opened new views of space and time; in the universe below and inside us, biology opened first glimpses of the foundation stones of experience: time, love, and memory.

What are the connections, the physical connections, between genes and behavior? What is the chain of reactions that leads from a single gene to a bark, or a laugh, or a song, or a thought, or a memory, or a glimpse of red, or a turn toward a light, or a raised hand, or a raised wing? The first scientists to look seriously at this question were the revolutionaries who figured out what genes are made of atom by atom—the founders of the science now known as molecular biology. Seymour Benzer was one of those revolutionaries, and he and his students took the enterprise farthest. Benzer's work on the problem was quiet, his students' work was quiet, and their story has never been told. But to a large extent the hard science of genes and behavior came out of their fly bottles. In this sense the fly bottle is one of the most significant legacies that the science of the twentieth century bequeaths to the twenty-first, a great gift and disturbance that human knowledge conveys to the night thoughts and day-to-day life of the third millennium. Pascal quoted Saint Augustine: "The way in which minds are attached to bodies is beyond our understanding, and yet this is what we are."

FROM A SHELF in his workroom, Benzer takes down a dusty set of test tubes. They are bound together in such a way that he can slide one test tube mouth to mouth with another test tube, like one cup lidding another, to form a series of sealed glass tunnels. They look something like panpipes. The design is so simple that the first model he built back in the 1960s still works. Now the London Science Museum has a replica, and someone from the San Francisco Exploratorium wants to automate one so that it will cycle through its paces over and over inside a glass display case.

Benzer dusts off the test tubes and lays them down flat on the benchtop before him. Then he lays a dim fluorescent bulb of fifteen watts on the far side of the benchtop. When he switches off the overhead lights, the fluorescent bulb glints on the test tubes and gleams on his reading glasses. The rows of bottles and bins, the stacks of books and manuscripts drop halfway into shadow. The light just catches the outlines of an ammonite propped against the far wall, a fossil shell the shape of a coiled elephant's trunk; and a row of trilobites, stone fossils with bulbous eyes. In the far corner of the sanctum a human brain sits in the dark. Benzer keeps meaning to find a proper jar for the brain. He wants to put it on his desk as a *memento mori* or a *memento vivere* ("Remember to live"). The brain waits in a bucket of formaldehyde, and what is left of its spinal cord curls in the bottom of the bucket like the lifeline of an embryo.

Benzer got the idea for his panpipes one night in 1966, when he put two test tubes mouth to mouth to make one long tube with a single fruit fly trapped inside it. He turned out the light; he rapped the tubes on his benchtop to make the fly drop to the bottom; and he laid the tubes flat on the benchtop with the fly at one end of the tunnel and a small dim light at the other. Sitting in the shadows, he watched the fly in the tunnel move toward the light, just as he had expected it to do, because according to the textbooks a grown fruit fly in a dark place is attracted to light—so is a grown human being in a similar situation. The next fly also moved toward the light. But he was surprised to see that when he put a single fly through this simple trial a few times in a row, the fly did not always do the same thing. One fly raced to the light once, walked to it the next time, and then quit. Another fly ignored the light the first time and then raced for it at the next opportunity. Most flies did choose the light most of the time, but each trial seemed unpredictable.

In 1966 it was already clear that however else historians would remember the twentieth century, they would remember it for the discovery of the atomic theory of matter and the atomic theory of inheritance. Physicists and geneticists had developed both theories early in the century. At midcentury a small circle of young scientists, including Benzer and Francis Crick (both lapsed physicists) and James Watson (a lapsed ornithologist) had united the two theories. They discovered what genes are made of atom by atom—the double helix, the spiral staircase of DNA; they mapped the fine structure of the gene down to the level of its atoms; and they cracked the code in which the genetic messages are written. They now knew precisely what a gene is physically, although they did not know how to connect the details they were looking at, which were atomic, with the details of the living world that most interested them and interest all of us: hands, eyes, lips, thoughts, acts, behavior. Within ten years the physicists-turned-biologists had learned so much about genes that they had begun to look around and above the genes for new worlds to conquer. To the boldest, many worlds beckoned, innumerable lines of work radiated outward from the gene, including the problems of the origin of life; the growing embryo; consciousness; and behavior, a problem that Crick called "attractively mysterious, one of the last true secrets in biology."

Watson, Crick, Benzer, and their circle had arrived at the double helix by working with viruses and *E. coli* bacteria in petri dishes. But they knew that geneticists before them had worked out the atomic theory of inheritance using fruit flies in milk bottles. Benzer has a strong, somewhat sentimental sense of history, and it appealed to him to make the next great leap forward by going back. Fruit flies are bigger than bacteria, but they are still tiny. They are grains of sand with wings, small enough to crawl through the mesh of screen doors, almost as small as Pascal's mites, so small that Aristotle mistook them for gnats. Coming from physics and from *E. coli,* Benzer saw them as atoms of behavior, and he thought they might be the perfect creatures with which to found a new science, an atomic theory of behavior.

By chance, the very first published laboratory study of *Drosophila,* or fruit flies, a long-forgotten paper that he tracked down some time afterward, had been a report on the flies' behavior: their reactions to light, gravity, and mechanical stimulation. Even that first report, which appeared in 1905, had suggested that the flies' instinct for light is not simple. If their jar was sitting on the windowsill—a biologist at Harvard reported—most of the flies would come to rest on the sides of the jar with their heads pointed away from the sun. But turn the jar slightly, and nearly every fly instantly flew toward the window.

To Benzer, *Drosophila* looked like just the happy medium he was looking for. An *E. coli* bacterium is a single cell. In a sense, he could think of it as a nervous system with a single neuron. At birth, a human baby has about one hundred billion neurons, one for every star in the Milky Way. A fruit fly has about one hundred thousand neurons, so it is the geometric mean between the simplest and the most complicated nervous system we know. Likewise, the mass of a single *E. coli* bacterium is one ten-trillionth of a gram. The mass of a man is one hundred thousand grams. The fly is roughly the geometric mean, at two thousandths of a gram. And a bacterium has a generation time of a hundredth of a day, while a human being has a generation time of ten thousand days (ten thousand days, to pick a generous round number, before one human being produces another). A fly has a generation time of about ten days, again roughly a geometric mean between the two. Even the number of genes in the fly is a mean between bacteria and human beings. In very round numbers a bacterium has 4,000 genes, a human being has 70,000 genes, and a fly has 15,000 genes, which puts it once again between the simplest and the most complicated creatures we know on the planet.

BENZER MODELED his test-tube experiment after a laboratory routine that he had learned from a chemist. The chemist used a simple trick to separate two compounds that were mixed together. One of his compounds would dissolve slightly more easily in oil and the other more easily in water. So the chemist put his mixture in oil and water and shook it up. He let the oil and water separate, oil above and water below. Then he transferred the top layer to a new tube and the bottom layer to another tube. He added fresh oil and water and shook them up again. When he had done this enough times, he found that he had separated the two compounds. The tube of oil now had an almost pure sample of the compound that liked oil, the tube of water an almost pure sample of the compound that liked water. Chemists call this the countercurrent distribution method because in a sense it sets currents flowing in opposite directions: one compound flowing upward, the other flowing downward.

So Benzer decided to make his own countercurrent apparatus. He assumed that most of the flies in the world's fly bottles like the light somewhat more than the dark but that a few might like the dark somewhat more than the light. He wanted to let the flies sort themselves into two more or less pure sets of particles, the light-lovers and the dark-lovers. Then he would look for the genes that made the difference. After some trial and error he hit on the idea of the panpipes. By mounting a set of test tubes in such a way that they could slide against each other, he could carry out a series of simple sorting operations, just like the chemist's.

Sitting now in the half-dark of his workroom, he uncorks one of his antique milk bottles ("Just a Little Better"), and he inserts a few dozen fruit flies into the tube on the far left of his countercurrent apparatus: Tube Zero. Then he raps the whole set of tubes on the benchtop a few times. In the quiet of his workroom in the middle of the night, the sound is like a pounding on the door. The raps knock the flies to the bottom of Tube Zero, and the flies swarm there a moment in the half-light, flailing as if in free fall. They are so small that they really do look like the Greek idea of atoms, points whirling in space, almost invisible and absolutely indivisible. (*Atomos* means "unsplittable.")

Benzer lays the countercurrent machine back down on the benchtop. The flies at the bottom of Tube Zero are now at one end of a glass tunnel. The only light they can see is the light at the far end of the tunnel, the fifteen-watt fluorescent tube. The flies can stay put in the bottom of Tube Zero, or they can move toward the light. As Pascal says, "There is enough light for those who desire only to see, and enough darkness for those of a contrary disposition." So the flies are facing a simple choice. If they do not move forward, they will remain in Tube Zero. If they do move forward, they will find themselves in the next test tube, Tube One.

Some of the flies walk, some of them run, some of them fly, and some of them meander. By the time fifteen seconds have passed, all but two of the flies have gone toward the light.

Benzer picks up the apparatus and slides the test tubes around. Now the flies that went toward the light—almost all of the winged atoms in the machine—are in Tube One, while the two flies that did not go remain in Tube Zero.

"Everybody gets another chance," Benzer murmurs. Again he raps the apparatus on the benchtop. Again he lays it flat on the benchtop. Within fifteen seconds, most of the flies again go toward the light. But this time one of the flies in Tube Zero decides to go too, and a few of the flies that went the first time choose not to go.

"What are they doing now?" Benzer says. "They're wandering around." This is why he built the countercurrent machine in the first place. Facing the same choice point, the flies do not always make the same choice. As a swarm, as a mass, they are predictable, but as individuals they are unpredictable. Even fruit flies in a test tube do not always act the same way twice. Why not? Watching their decisions and revisions by the light of the fifteen-watt bulb back in 1966, Benzer got an inkling that flies might be more than atoms of behavior. He had thought they would be simple, regular, predictable, like the particles in the chemist's test tubes. Instead the flies act as if they are improvising from moment to moment, based on what each fly sees in front of it, what has happened in its past, and what can only be called the personality of that fly.

"Flip it over, switch it back." Now the flies that have chosen the light twice are in Tube Two. The flies that have chosen the light once are in Tube One. The fly that has never chosen the light is still in Tube Zero, moving erratically in the half-dark. To human eyes it looks like a troubled fly.

"Now we'll give them *another* chance." Benzer rocks and slides his gadget: the same operation, the same fifteen-second wait. Then again and again: *knock, knock, knock.* Each time, most of the flies move toward the light.

"OK, there we are," Benzer says at last, switching on the overhead lights. He can read the results at a glance. Tube Six contains flies that moved to the light six out of six times. Most of them are in there. Tube Five contains flies that moved five out of six times. And so on, all the way down to Tube Zero, the home of the troubled fly, which may be damaged goods.

To prove that he is really testing what he thinks he is testing, Benzer dumps all of the flies back into Tube Zero and sets the apparatus on the benchtop again. This time he sets it down with the flies next to the light. Now the flies face the opposite choice. They are near the light. They can choose to stay where they are, or they can move away. When Benzer runs the experiment this way, most of the flies stay put in the bottom of Tube Zero. They do not hurry into the darkness the way they hurried into the light. Running the experiment backward this way is Benzer's control. It proves that in the first rounds the flies were not indifferent to the light; they were not advancing through the tubes just for the sake of advancing through the tubes. He is looking at what he thinks he is looking at. The light is the key.

PHILOSOPHERS used to speak of the ultimities. The last stop of a trip, the highest high C of a musical scale, and the final stage of an alchemical process, when the liquid in a beaker had traveled "from Crudity to Perfect Concoction," as Sir Francis Bacon put it—these were ultimities.

In science, the ultimities are the ultimate questions. These are the questions that so many generations have raised that they have come to seem eternal, always to be asked and never to be answered. They are the problems that interest us so much that solving them—finding even a small piece of them—would feel like finding the secret of life. The origin of species was once one of the ultimities of science, until Darwin. The origin of the universe is one of the ultimities today. So is the origin of life. And the most intimate, the most immediate, in some ways the most intricate and the most important for our inquiring species, will always be the origin of behavior. We have asked these questions from the beginning: How much of our fate is decided before we are born? What is written and in what code and of what materials? What are the connections between atoms, thoughts, feelings, behavior? How much of our behavior is passed down from one generation to the next?

Benzer's countercurrent machine was a point of origin for the science of genes and behavior, a point of origin for the headlines that have punctuated the news for the last ten years and sometimes seem likely to dominate it within another ten years. It was the beginning of what Benzer calls the genetic dissection of behavior.

This is a science that is dedicated to exploring the inward infinity that Pascal imagined and to reading the writing on John Locke's slate—for even Locke knew that the slate is not blank. He thought that our temperaments are at least partly innate: "Some men by unalterable frame of their constitution are stout, others timorous, some confident, others modest and tractable." He did not think that very much else about our minds is innate, although other eighteenth-century philosophers argued that we are born knowing a great deal: "that sweetness is not bitterness," to give one of the examples that Locke cited; or "that 'two bodies cannot be in the same place,' and that 'it is impossible for the same thing to be and not to be,' that 'white is not black,' that a 'square is not a circle,' that 'yellowness is not sweetness.' " Many philosophers assumed that "these and a million of other such propositions," as Locke skeptically wrote, "must be innate." Today, these are questions that the science of genes and behavior can begin to test at the level of the hipbone-is-connected-to-the-thighbone.

Sigmund Freud tried to make a solid science of human behavior. "Have you not noticed," he wrote early in the century, "that every philosopher, every imaginative writer, every historian and every biographer makes up his own psychology for himself, brings forward his own particular hypotheses concerning the interconnections and aims of mental acts—all more or less plausible and all equally untrustworthy? There is an evident lack of any common foundation." Freud tried to establish a foundation as solid as the foundations in physics and chemistry; but today the most interesting effort in progress is founded on physics, chemistry, and Benzer's beginning.

The new effort also builds on Darwin, and on the Darwinian studies of behavior that were attempted in the 1930s and 1940s by Konrad Lorenz, Niko Tinbergen, Karl von Frisch, and their students, who called themselves ethologists. One of Tinbergen's books is illustrated by the silhouette of a bird in flight. When newborn goslings see that silhouette in the sky, they read the shape as a goose if it is moving to the left, a hawk if it is moving to the right. The silhouette of the goose does not scare goslings, but the silhouette of the hawk sends them scurrying. That kind of hard fact fascinated the ethologists. Goslings don't learn to make that distinction between friend and enemy from their mothers. They know it from the first moment they see the sky. They know it when they are still standing in the nest with caps of eggshell on their heads. How do they know? Ethologists looked at such pieces of behavior and tried to dissect them into routines and subroutines, which they called "atoms of behavior." Now with the tools of genetic dissection biologists can actually begin to study the instincts of goslings and newborn babies at the level of the atoms.

In the 1970s, E. O. Wilson, inspired in part by his studies of ant societies, tried to extend the work of the ethologists to human beings in a new synthesis he called "sociobiology." Wilson was attacked by social scientists, on the one side, because they hated his attempt to biologize human nature, and he was attacked by biologists, on the other side, because they felt he had built castles in the air and had not acknowledged the need for hard molecular biologists to put foundations under them. Today many of those sociobiological speculations can be explored in the code of the ants, the code of the flies, and the code of human beings; and Wilson himself wants to claim the genetic dissection of

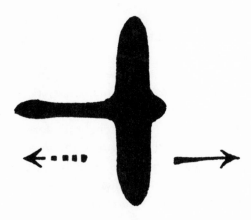

*What are we born knowing? This shape is often the first thing that newly hatched goslings see in the sky. If the shape is gliding to the left, the goslings stay in their nest; to the right, they scatter. Leftward means safety (goose); rightward means danger (hawk). All animals—including the human animal—inherit pieces of behavior. Biologists are now beginning to dissect some of our oldest and deepest instincts at the level of the genes.*

behavior as a cornerstone of his foundation. "Better Benzer than Freud! Quote me. Better Benzer than Freud!" Wilson says, standing in his office by one of his celebrated colonies of leaf-cutter ants, to which he feeds *Drosophila*—preferably wingless mutants.

"THERE IS GRANDEUR in this view of life," Charles Darwin writes in the last lines of *The Origin of Species*. Darwin found grandeur in the thought that life, "having been originally breathed into a few forms or into one," has gone and is still going in so many astonishing directions; that life, "whilst this planet has gone cycling on according to the fixed law of gravity," has produced and is now producing so many astonishing species and varieties, from viruses and bacteria to grass, from oaks to peacocks, great apes, and great whales; that "from so simple a beginning endless forms most beautiful and most wonderful have been, and are being, evolved."

Of all the endless forms that have been and are being evolved on this planet, some of the most wonderful (if not the most beautiful) have emerged from the laboratories of Seymour Benzer and his school. The science they make possible is changing our view of life; and both the most passionate proponents and opponents of the science believe that in the twenty-first century it may change the conditions and prospects of life. Many compare this moment to the sudden acceleration of science itself in the seventeenth century, Pascal's century, or of atomic physics in the first half of the twentieth century. Others say the pace is unprecedented. "To compare the speed with which understanding is being deepened in the life sciences with what happened in physics in the 1920s is probably flattering to physics," the former editor of the journal *Nature* wrote not long ago in his farewell editorial, after several decades of watching the molecular revolution. "Can there ever have been a time when there have been so many people pushing at an open door?"

As the twentieth century ends, more and more of the best and brightest in science are being drawn to the effort. Molecular biology has made biologists the royalty of science. James Watson has installed one of the science's golden boys—one of Benzer's best students' best students—next to his own office at Cold Spring Harbor Laboratory, New York, where Watson is the director. Sometimes late in the evening Watson wanders into the Fly Rooms and looks at the madding crowds in the tidy bottles. He has a nimbus of white hair, but he is still almost as gaunt and gangly as he was at the age of twenty-four, when he cleared a space on his desktop, slid together a few puzzle pieces of tin and cardboard, and saw the double helix. Now when he looks at the mutants in the Fly Rooms next to his office, he feels that he is looking at the beginning of the twentieth century and also the beginning of the twenty-first. This is where his science started, and this is where it is going. "And—I just find it very—" Watson says with a blue-eyed cosmic stare. "You know, just divorcing yourself from humans—just forgetting about us. That there are certain complicated behaviors that seem to be inherited. That's really—" He laughs through his teeth with what sounds like a snicker of glee, as he must have laughed four decades ago when he was talking with Crick about the physical structure of the gene. "You know," he says, and he snickers again, *"that's the problem to solve."*

In the study of the physical links between genes and behavior, many of the first answers and half answers came from flies: time, love, memory, as viewed through a compound eye. In one way, this is a parable of the strangeness of life: so much millennial science from such tiny and alien creatures. In another way, it is a parable of the unity of life, since not only flies and human beings but everything alive springs from even homelier materials: the same genes, the same atoms, the same clay, the same simple beginnings. In the first view, the whole world looks alien to us; in the second view, there is nothing on earth that is not familiar.

Knowing what he does now, Benzer would not dream of calling a fruit fly just an atom of behavior. His research has reached that penultimate stage during which he would rather work than eat or sleep. In the middle of the night, after putting his test tubes back on their shelf, he unhoods a microscope and examines the head of a fly. "About the size of the head of a pin," he says. "A hundred thousand angels on it. Dancing."

# CHAPTER TWO

## The White-Eyed Fly

*Those who love wisdom must be inquirers into many things indeed.*

—HERACLITUS

THE QUANTUM PHYSICIST Richard Feynman once gave a lecture on color vision in Caltech's Beckman Auditorium. He explained the molecular events that take place in the human eye and brain to show us red, yellow, green, violet, indigo, and blue. This chain of reactions was one of the early discoveries of molecular biology, and it fascinated Feynman. "Yeah," someone in the audience said, "but what is really happening in the mind when you see the color red?" And Feynman replied, "We scientists have a way of dealing with such problems. We ignore them, temporarily."

That line still makes Benzer smile in the middle of the night, which is the middle of his working day. He often repeats it to the postdoctoral students in his laboratory. *"We ignore them, pause, temporarily.* I thought that was a wonderful statement," he says. "You know, we do that all the time. The problem you have an instinct for, a feeling for, may be a problem that you're not going to be able to solve, and you sort of shy away from it, temporarily. The philosophers, of course, don't ignore them. Maybe they have a penchant for unsolvable problems. If it's solvable, it's not really interesting."

The problems that Benzer and his students are solving are problems that scientists and philosophers from the beginning of recorded history have been unable to solve and also unable to ignore. In the fifth century B.C. Hippocrates, the great-grandfather of modern medicine, dropped in on Democritus, the great-grandfather of atomic physics. He found Democritus sitting in a garden with the bodies of dead beasts cut open all around him. Democritus was sleeping and waking over his notes, trying to find the cause of melancholy, so that he could cure it in himself and teach others to cure it too. It was a first attempt at the dissection of behavior, approximately twenty-five centuries too soon.

In the second century A.D. the Greek physician Galen reported that he and a few of his friends had delivered a young goat by cesarean section "so that it would never see the one who bore it." They took the kid from its mother's womb and placed it in a room in which there were bowls of wine, oil, honey, milk, grains, and fruits. "We observed that kid take its first steps as if it were hearing that it had legs," Galen wrote; "then, it shook off the moisture from its mother; the third thing it did was to scratch its side with its foot; next we saw it sniff each of the bowls in the room, and then from among all of these, it smelled the milk and lapped it up. And with this everyone gave a yell, seeing realized what Hippocrates had said: 'The natures of animals are untutored.' "

Hippocrates and Galen tried to relate human temperaments to the elements: fire, air, earth, and water. The word temperament comes from the Latin *temperare,* "to mix"; in Galen's scheme every human being is a mix of those elements. Astrologers tried to relate temperaments to the stars, looking for connections between the two infinites, the skies over our heads and the skies inside our heads. Some of the symbols on fly bottles are astrological. Virgo is one of the twelve signs of the zodiac, part letter, part hieroglyph, part Hebrew, part Phoenician, and on a fly bottle it means what it has always meant, *Virgin.* The circle with the pendant cross, meaning *Female,* was once the sign for the planet Venus, with connotations of fertility. The circle with the angry arrow, meaning *Male,* was once the sign for the planet Mars, with connotations of calamity.

"What a wonderful thing it is that that drop of seed from which we are produced bears in itself the impressions, not only of the bodily shape, but of the thoughts and inclinations of our fathers!" Montaigne wrote in the sixteenth century. "Where can that drop of fluid harbor such an infinite number of forms? And how do they convey those resemblances, so heedless and irregular in their progress, that the great-grandson shall be like his great-grandfather, the nephew like his uncle?" The questions were just as unanswerable in Montaigne's century as in Galen's.

Shakespeare seems to have been the first to use the words "nature" and "nurture" in brooding about these mysteries. In his last play, *The Tempest,* which he completed in 1612, Prospero (the character who comes closest to a self-portrait in any of Shakespeare's plays; the archetype of all artists, scientists, and philosophers) complains about his adopted son, Caliban:

> *A devil, a born devil, on whose nature*
> *Nurture can never stick; on whom my pains*
> *Humanely taken, all, all lost, quite lost.*

In one way or another, the paradoxes of nature and nurture fascinated every poet, every playwright, and every pair of parents from the first. Abel was a keeper of sheep, but Cain was a tiller of the ground. Yet they both sprang from Adam and Eve. Esau was a cunning hunter, a man of the field; but Jacob was a plain man, dwelling in tents. Esau was also a hairy man, but Jacob was a smooth man. Yet Jacob followed Esau out of the womb with his hand on Esau's heel.

In the late 1830s, when Charles Darwin first began scribbling a theory of evolution, his handwriting in his secret notebooks reminded him of the handwriting of his old radical grandfather Erasmus, who had published a theory of evolution the century before. Darwin wondered if he was looking at an "instance of hereditary mind." His cousin Francis Galton later closed his copy of the *Origin* wondering if he had just devoured the book because of a "bent of mind that both its illustrious author and myself have inherited from our common grandfather, Dr. Erasmus Darwin." Darwin and Galton each spent decades compiling examples and anecdotes of what they vaguely called the power of inheritance— vaguely, because even though evolution depended on it, they had no idea how the power of inheritance actually works. They never got much closer to that secret than Galen had, although Darwin's theory framed the question for those who came after him like the framing of a doorway.

Until the twentieth century, the passage of time made less difference with this problem than with almost any other basic problem in the study of life. Every argument about nature and nurture was like the mutant fly *pirouette,* spiraling inward until it died. Every new system of thought that claimed to solve the problem floated apart from every other, cut off from the rest of human knowledge like the brain in Benzer's workroom, with its severed cord drifting in formaldehyde. The problem was unsolvable before the discovery of the gene.

THAT DISCOVERY was made by a monk in what is now Brno, in the Czech Republic, in a report on garden peas. Gregor Mendel tended a monastery garden and an experimental greenhouse in the 1850s. While crossing different strains of peas two by two—dusting the flowers of one strain with the pollen of the other—Mendel got a clearer view of the patterns of inheritance than anyone before him. The strains he crossed had smooth peas or wrinkled peas, yellow peas or green; their plant stems were tall or short; and so on: seven pairs of contrasting strains. When Mendel crossed these strains, the traits did not blend together but passed on intact, often skipping several generations. This experiment was so simple that it could have been done by Hippocrates or Democritus, and what made it revolutionary was its demonstration that inheritance comes in something like the Greek idea of atoms. Tallness never blended with shortness. They do blend in humans, but they don't blend in peas—one reason Mendel was lucky to work with peas. There the patterns stayed crisp and clear. The traits stayed separate generation after generation, and Mendel assumed that they must be governed by separate factors. Much later, biologists who reread Mendel's paper would think of these factors, with a nod to physics and chemistry, as particles of inheritance. No one knows how the monk pictured them, although he, like Benzer, had been trained as a physicist.

Thirty years ago, Benzer and another founder of molecular biology, Gunther Stent, climbed Mount Fuji, a climb that Benzer still remembers fondly because on it he tasted his first mandarin orange. They stopped for the night in a Buddhist temple halfway up the volcano. The temple's caretaker was an ancient woman, who asked Benzer and Stent, through their interpreter, what they did. As they began to explain, she said, "Ah, Mendel."

While Mendel planted peas, Francis Galton, Darwin's cousin, caught glimpses of the particles of inheritance in human beings. Reading the *Origin* had inspired Galton to send out hundreds of letters and questionnaires to friends, acquaintances, acquaintances of acquaintances, asking about family resemblances and especially about twins. Galton spent the rest of his life collecting such data and inventing new statistical tools to analyze the patterns. Along the way he sent Darwin an example of behavior that skipped generations:

> A gentleman of considerable position was found by his wife to have the curious trick, when he lay fast asleep on his back in bed, of raising his right arm slowly in front of his face, up to his forehead, and then dropping it with a jerk, so that the wrist fell heavily on the bridge of his nose. The trick did not occur every night, but occasionally, and

was independent of any ascertained cause. Sometimes it was repeated incessantly for an hour or more. The gentleman's nose was prominent, and its bridge often became sore from the blows which it received. . . .

Many years after his death, his son married a lady who had never heard of the family incident. She, however, observed precisely the same peculiarity in her husband; but his nose, from not being particularly prominent, has never as yet suffered from the blows. . . .

One of his children, a girl, has inherited the same trick.

"We seem to inherit bit by bit," Galton concluded in 1889 in his book *Natural Inheritance*. That was the only way Galton could interpret the peculiar persistence of bits of family resemblance and bits of behavior. But the patterns of inheritance in his data were not as clean and clear-cut as they were in Mendel's experiments; and neither Galton nor Darwin ever read his paper, which was published in the journal of Mendel's local natural history society in 1866.

No one realized what Mendel's paper might mean for the inheritance problem until a botanist cited it in a paper in January 1900. By then the time was right, and two other citations followed that same year. All three papers attracted attention, although the existence of atomic particles was still considered speculative, and so was the existence of Mendel's particles of inheritance. One of the biologists who read the new papers closely but skeptically was Thomas Hunt Morgan, born in Lexington, Kentucky, in 1866, the same year that Mendel had published his paper.

In the fall of 1907, Morgan, then a professor of zoology at Columbia University, told one of his students to put a few bananas on the ledge of his laboratory window to attract some fruit flies. Neither teacher nor student was thinking about Mendel at the time; the student wanted to breed animals in the dark and see if they would lose their instinct to go toward light. Morgan told him to use fruit flies because his laboratory in Schermerhorn Hall was cramped, only about sixteen by twenty-three feet. It was cramped and crowded because Morgan, who had been trained as an old-fashioned naturalist, already took pleasure in working with pigeons, chickens, starfish, rats, and yellow mice.

So Morgan's student Fernandus Payne trapped some flies, put them in darkness, and let them breed. Within a short time he thought he could detect a change. The tenth generation seemed to move toward the light a little more slowly than the first. This student project, which Payne wrote up in a paper, "Forty-nine Generations in the Dark," was soon lost in the late prehistory of genetics. Benzer knew nothing about it in 1966 when he built his countercurrent machine.

Next Morgan decided to see if he could force fruit flies to change faster. Morgan had money, but he was a miser in the laboratory, which was another reason he liked working with flies. For microscope lamps, he used ordinary lightbulbs with shades that he and Payne cut out of tin cans. For fly bottles, according to legend, he and his students stole empty half-pints from milk boxes on Manhattan stoops on their early-morning walks to the lab, and they lifted more from the Columbia student cafeteria. This was the milk-bottle tradition that Benzer would inherit.

Morgan subjected his flies to heat, cold, and X rays, trying to create a fly that looked different in some way from all the other flies. He also injected the flies' private parts with acids, bases, salts, sugars, and alcohol. Beneath the hand lens each fly had the same six legs, the same veined and cross-veined wings, the same brilliant red eyes; but Morgan kept watching and waiting for a mutant. And this was the enterprise that Benzer would inherit, although Benzer would manipulate his flies with more sophistication, and he would watch for changes not in their bodies but in their behavior.

Early in the fall of 1909, Morgan began trying to speed up the evolution of his flies with a new approach. He focused on a dark pattern on the thorax of the fly, a variable pattern in the shape of a trident. Week after week he bred only those flies with the most variable tridents and waited to see if the pressure of this artificial selection would somehow set off an explosion of mutations in one of his fly bottles. Week after week, he and Payne saw nothing but ordinary tridents on ordinary flies.

"There's two years' work wasted," Morgan told a visitor in the first days of January 1910, waving at shelves of stolen milk bottles. But just a few days later, Morgan found a fly with a trident that was slightly darker and more sharply defined than before. Then he found a fly that had a dark blotch where the wing met the thorax. Then, after all those tens of thousands of more or less identical red-eyed flies, he found a single fly with white eyes.

Morgan's wife, Lilian, who was fascinated by his work and who later (after their children were out of the house and busy in school) made important contributions in the laboratory, was pregnant that year; and long afterward the birth of the new baby became mingled in the family history with the arrival of the mutant. Lilian loved to recall the scene when Morgan walked into her hospital room.

"Well, how is the white-eyed fly?" she asked. According to family lore, he was carrying the fly home at night to sleep in a jar next to their bed.

Morgan told her the fly looked feeble but it was hanging on. "And how is the baby?"

Within a week, one of their two new arrivals was old enough to breed (still another reason to work with flies). Morgan paired the white-eyed fly, which was a male, with normal virgin female flies, and together they produced 1,237 young flies. The flies' children (as Morgan called them) had red eyes. The next week, Morgan arranged marriages for the children. He was fascinated to see that among the grandchildren, although all of the females had red eyes, about one in two of the males had white eyes. Naturally Morgan thought of Mendel's peas. When Mendel crossed short peas with tall peas, the first generation was all tall, and in the next generation three quarters of the plants were tall and one quarter was short. Shortness in Mendel's pea plants is what is now known as a recessive trait, like blue eyes among human beings. Morgan wondered if white eyes among male fruit flies could be a recessive trait too.

As one latter-day drosophilist likes to say now, "In the beginning there was *white*." The mutant fly *white* was the point of entry through which Morgan would establish the modern theory of the gene, the atomic theory of inheritance. In much the same way, decades later, the arrival of the first clock mutant, a fly without a sense of time, would open the atomic theory of behavior. That mutant would give Benzer a point of entry through which he and his students could begin the remarkable series of experiments in which they took basic instincts apart and put them back together, like clock makers who had unscrewed the back of a clock.

The arrival of *white* intrigued Morgan because ever since 1900, biologists had been looking for Mendel's elements. Through microscopes they could see tiny threads called chromosomes in the nucleus of every egg. They could also see that when a spermatozoan enters an egg, it contributes a matching set of threads. To many biologists, this looked like a physical explanation for Mendel's results in his pea garden. A pea plant might inherit a tallness factor on one chromosome, for example, and a shortness factor on the other chromosome in that particular pair. The logic seemed compelling: Mendel saw traits in pairs; microscopists saw chromosomes in pairs. And when a young body begins making new eggs or new sperm, each egg cell and sperm cell receives only one chromosome from each chromosome pair. That way, when sperm finds egg, the single chromosomes can meet inside the fertilized egg and the whole process can start over to make a new life.

All of this fit Mendel's observations. But Morgan was a contrarian. He used to complain that his best friends at Columbia were "wild over chromosomes." It did look as if chromosomes were *the thing*," he said, but he wanted hard evidence: "I cannot but fear that we are rapidly developing a sort of Mendelian ritual by which to explain the extraordinary facts." Morgan's "show-me" attitude was about to lead him into the simple line of experiments that started the revolution in twentieth-century biology. It was the same attitude that Benzer would later bring to the study of genes and behavior in his own Fly Room. Feynman, the quantum physicist, once knocked on the door of Benzer's workroom and asked him to show his son a fly's brain. Benzer sat the boy at a microscope and told him, "There're a hundred thousand transistors in that brain." Benzer's own work as a physicist had helped lead to the invention of the transistor. By comparing the fly's one hundred thousand neurons to transistors, Benzer was trying to convey an idea of the fly brain's magnificent miniaturization. At the same time Benzer was also nodding to the father over the son's head, physicist to physicist.

But Feynman said, "No, no. Tell it straight. They're not transistors, they're neurons. Don't oversimplify."

Benzer liked that line too. Feynman was right. A neuron is actually a much more complicated object than a transistor, and the path from gene to neuron and from neuron to behavior is longer and more mysterious than the path from an electron to a radio or a computer. Benzer shared with Feynman an aggressively simple and direct style of talking—a trait both men also shared with T. H. Morgan. Benzer and Feynman came from families of New York Jews, Eastern European immigrants; Morgan came from an old family of Kentucky aristocrats. But all three spoke in the kind of down-home, common-as-flies style that is the lingua franca of great scientists, conveying a contempt for pretension, a contempt for cant, a delight in common sense, combined with uncommon curiosity about what is really there.

Sitting in his first Fly Room in Columbia's Schermerhorn Hall, T. H. Morgan could see through his microscope that a fruit fly has four pairs of chromosomes. In female flies, all four pairs look alike: short, featureless threads. But in male flies, the fourth pair looks different: one is bigger than the other. This is the chromosome pair that is now known, famously, as the X and the Y. Morgan focused on this mismatched fourth pair. He knew that a fly, like a pea plant or a human being, always inherits one chromosome from each parent. In each pair, one chromosome comes from the father and one from the mother. Since a female fly has two Xs

*In his Fly Room at Columbia University, in the early years of the twenti-
eth century, Thomas Hunt Morgan transformed the study of life.
Because Morgan hated being photographed, his students stole this picture by hiding the camera in an incubator and pulling a string. The camera, along with the books and microscopes that are visible behind
Morgan, belonged to his favorite student, Alfred Sturtevant, whose
eureka one night in 1911 helped establish the theory of the gene.*

and a male has an X and a Y, Morgan could deduce that a son must inherit
his X from his mother and his Y from his father. If his father has white eyes
and his mother has red eyes, then he will have red eyes. But if his father
has red eyes and his mother has white, then he will have white eyes. So
Morgan wondered if the fly has a gene for eye color on the X chromosome.

A female fly gets one X chromosome from each of her parents. If her mother gives her an X with a gene for white eyes and her father also gives her an X with a gene for white eyes, then she will have white eyes. But if either of her parents gives her an X with a gene for red eyes, then she will have red eyes, because red is dominant over white in flies, just as purple is dominant over white in the flowers of Mendel's pea plants.

Human beings have many more pairs of chromosomes (twenty-three pairs, although Morgan's generation could not sort them out and count them properly). Twenty-two of those twenty-three pairs look like identical twin threads under the microscope. The last pair is mismatched, just as it is in flies: two X chromosomes in women, an X and a Y in men. And color blindness in men, Morgan realized, as he thought about all this, "follows the same scheme as does white eyes in my flies."

Morgan began to suspect that genes might exist, and that there really might be a gene for eye color hidden somewhere on the fly's X chromosome. By now he and his first students had examined so many flies through their hand lenses and microscopes that the slightest aberration leaped out, and they began to find more and more mutants in their fly bottles. One was a fly with abnormally short wings. When Morgan bred it he saw that wing length, like eye color, seemed to be on the X.

Now Morgan experimented with these mutants. Suppose a female has red eyes and long wings: she is a normal fly. Suppose a male has white eyes and short wings: he is a double mutant. If they mate, every one of their daughters will inherit one X from the mother and one X from the father. So, since red and long are dominant over white and short, the daughters should have red eyes and long wings, too. Morgan arranged that cross, and so it was.

Morgan thought he knew exactly what was on each of those two Xs, if the gene theory was correct. One X carried genes for red eyes and long wings; the other X carried genes for white eyes and short wings. If so, then when these normal-looking females mated with normal males, each mother should produce just two kinds of sons, depending on which X each son inherited. Some of the sons should be normal like their mother; some of the sons should be double mutants like their grandfather.

Morgan arranged this cross. Just as he expected, some of the sons were normal and some were double mutants. But others had white eyes and normal wings, and still others had red eyes and short wings: they were single mutants. At first sight, that seemed impossible. Each of the sons could inherit just one X from his mother, and his mother did not have an X with a white gene and a long gene, or an X with a red gene and a short gene. It was as if the mother had mixed and matched bits of her two Xs before passing out an X to each of her sons.

After much thought, Morgan could explain that. He considered the microscopic action that takes place when a female fly makes an egg. Her egg has to receive four chromosomes, one strand apiece from each of her four pairs of chromosomes. But the specialized cell that prepares the chromosomes for the new egg has the chromosomes in pairs. So to produce the egg, in the process known as meiosis, each of those pairs of chromosomes has to split up. Just before they part, each pair does something almost gaudily bizarre. The two strands twist and twine around each other as if they themselves are mating. They writhe together like copulating snakes.

So Morgan imagined two of these X chromosomes mating, so to speak—wrapping around each other, aligning themselves so that every point on one X touches the corresponding point on the other. During that intimate moment, Morgan thought, bits of each chromosome might somehow trade places. Genes might cross over from one X to the other. Afterward, the solitary X chromosome that passes into the female fly's egg would carry some genes from her father's X and some genes from her mother's X. And all this shuffling and crossing-over might have caused the oddities that Morgan was trying to explain. The genes on the X chromosomes had shuffled; they had mixed and matched.

Morgan arrived at this vision of crossing-over toward the end of 1911. In the Fly Room, he shared it with his favorite student, Alfred Sturtevant, then a senior at Columbia. What happened next is one of the most important eurekas in twentieth-century science and deserves to be better known outside the field of genetics. The moment would help to define both the style and substance of the study of life for the rest of the century.

This crossing-over idea had interesting implications, Morgan told Sturtevant. Picture that female fly and her two Xs just before the shuffle and scuffle of crossing-over. On one X, she has genes for red eyes and long wings. On the other X, she has genes for white eyes and short wings. Suppose these two genes lie very close together on the X. If they are close, then during crossing-over the two genes will be likely to stay together, like two people who are standing right next to each other in a swirling crowd. But suppose the two genes lie at opposite ends of the X chromosome. Then they will have more chance of being separated, like two people who are standing farther apart.

Morgan thought he could apply this idea to the results of their breeding experiment. If single mutants were rare, that would imply that the gene for eye color and the gene for wing length must lie close together on the X, because the two genes had not been separated very often during crossing-over. But if single mutants were common, that would imply that the two genes must lie far apart on the X, because they had been separated so often. And in fact, the experiment had produced quite a few single mutants: about 30 percent of the sons were single mutants.

Sturtevant not only followed all this: there in the Fly Room, as he listened to Morgan, he had the idea of his life. By this time, he and Morgan and the other students in the Fly Room had found quite a few genes that seemed to lie on the fly's X chromosome. Sturtevant realized that if Morgan was right about crossing-over, he might actually be able to figure out where each one of these genes lies on the X. He could test Mendel, he could test Morgan, and he could make a map of genes on a chromosome, all in one stroke.

That afternoon Sturtevant collected a stack of laboratory records, the complete records of crosses involving half a dozen genes, and he took the papers home. At home he spread out the papers. He imagined a half-dozen beads on a string, or points on a line:

$$\underline{\overset{\bullet}{A} \quad \overset{\bullet}{B} \quad \overset{\bullet}{C} \quad \overset{\bullet}{D} \quad \overset{\bullet}{E} \quad \overset{\bullet}{F}}$$

If genes are real and if they lie in a straight line on a chromosome, they must be in this linear order. *A* must be closer to *B* than to *C* and so on. So in each generation of flies, there should be more *AB*s than *AC*s, because *A* and *B* are more likely to travel together than *A* and *C*. By now Morgan and his students had crossed tens of thousands of flies and they had kept records of each cross. So Sturtevant checked to see which genes had stayed together more often and which genes had parted more often when the flies were crossed.

In the beginning there was *white*. Which genes are close to *white*? Sturtevant thought he could guess one of them. In the Fly Room they had found a gene that affects the body color of the fly. They called this gene *yellow* because they had first inferred its existence when they came across a yellow-bodied mutant. For his purposes now, Sturtevant needed to see the results of a cross in which one parent had white eyes and a normal brown body, and the other parent had red eyes and a yellow body. If the eye-color gene and the body-color gene were very close together, virtually all of the descendants should be of two kinds: either white-eyed

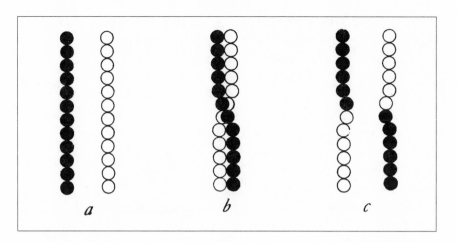

*Proof that genes are real. This is T. H. Morgan's own diagram of the phenomenon known as crossing-over, which one biologist has called "arguably the most intimate event in sexual reproduction." (a) A schematic drawing of a single pair of chromosomes. Morgan pictures the genes on the chromosomes as pearls on strings. The black pearls come from the mother, the white from the father. (b) Just before the creation of an egg cell, the two chromosomes twine together, and genes cross over. (c) Now each chromosome in the pair carries some genes from the mother and some genes from the father. Morgan and his students used crossing-over to make the first genetic maps, in one of the most extraordinary series of experiments in the twentieth century.*

and brown-bodied, or else red-eyed and yellow-bodied. But if the genes were farther apart, they would be separated more often during crossing-over. In that case, many of the two flies' descendants would have white eyes and yellow bodies.

Sturtevant checked the breeding records he had brought home. In the records, exactly 21,736 flies were the descendants of such parents. Out of those 21,736 fly children, only 214, or about 1 percent, had white eyes and yellow bodies. So those two genes had rarely been separated during crossing-over. That meant the *yellow* gene must lie very close to *white*.

Sturtevant decided to call 1 percent one map unit. He would say that *white* and *yellow* are one map unit apart.

The lab records he had brought home also included a cross between flies with yellow bodies and flies with vermilion eyes. Those crosses had produced 4,551 fly children. Of those children, 1,464 flies, about 32 percent, had inherited both yellow bodies and vermilion eyes. If 1 percent is

*"Who could have foreseen such a deluge?" Morgan wrote when flies and fly bottles began crowding out everything else in his lab at Columbia. The science was glamorous; the Fly Room was anything but. Morgan and his students arrived at the lab each day bearing more and more empty half-pint milk bottles, which they stole from Manhattan stoops and from the Columbia student cafeteria. Note the bananas hanging in the corner: food not only for the flies but also for the world's first geneticists. This photograph was taken around 1920.*

one map unit, 32 percent is thirty-two map units. So *yellow* is one map unit away from *white* and thirty-two map units away from *vermilion*.

Next came a third cross. In the records, there were 1,584 fly children that had one *white* parent and one *vermilion* parent. Of those 1,584 children, he found that 471, about 29 percent, had inherited both *white* and *vermilion*. So *white* was twenty-nine map units away from *vermilion*.

Now came the moment that mathematicians, when they describe a brilliant equation, call the beauty part. Generations of geneticists have since retraced Sturtevant's big night and shaken their heads over the simplicity of the trick that started their revolution. Much later Benzer would return to this trick, give it a twist, and start a second revolution.

Sturtevant was looking at a simple mathematical puzzle. He knew that *white* is closer to *yellow* than to *vermilion*. He knew that *yellow* is closer to *white* than to *vermilion*. And he also knew that the distance between *yellow* and *vermilion* is greater than the distance between *white* and *vermilion*. There is only one way to explain those numbers if the genes are on a straight line. They have to be arranged like this:

yellow white            vermilion

Sturtevant checked the numbers. He had one map unit between *yellow* and *white;* thirty map units between *white* and *vermilion;* thirty-two map units between *yellow* and *vermilion.* So far, then, everything seemed to be in order, "at least mathematically," as he wrote later. The arithmetic was close enough, given the slight fuzziness in the data. And when he checked the rest of the breeding data, all of the other numbers and distances fit too. He placed a mutation called *miniature wing* about three map units away from *vermilion.* The wings of *miniature* are normal in shape but very short, like human arms and hands that reach only to the belt. He placed *rudimentary wing* about twenty-four map units from *miniature.* The wings of *rudimentary* are a bit of a mess: Some are wrinkled and blistered; some are truncated; some have irregularly spaced hairs.

Before dawn, Sturtevant was finished. As a senior at Columbia, he had a full load of assignments from other courses, and he had just put in an all-nighter on a long-shot project that no one had assigned. "I had quite a lot of homework to do," Sturtevant used to say long afterward, "but I didn't do any of it; but I did come back with a map the next morning." In the Fly Room, he laid out the first genetic map, with the genes spaced out in a line:

yellow white            vermilion miniature          rudimentary

Looking at this simple map, Morgan and his students could see that what they had been supposing and assuming month after month in their flyspecked laboratory was almost certainly true. Genes are real, genes are on chromosomes, and genes can be surveyed and explored. It was the biggest lightning flash and thunderclap in biology since the rediscovery of Mendel in 1900. Morgan, who was not given to hyperbole, once called the view the map opened "one of the most amazing developments in the whole history of biology." Sturtevant was nineteen years old.

Morgan and his students would spend the next several decades mapping more and more genes on the X and the other three fly chromosomes, and convincing the doubters inside and outside biology that genes are real.

*Alfred Sturtevant, T. H. Morgan's favorite student, figured out how to draw a map of the genes in 1911, at the age of nineteen. His discovery started biology on a long march inward. Sturtevant made gene mapping his life's work; he never left Morgan. This photo was taken around 1925.*

Long afterward, when Benzer and his student Ronald J. Konopka found the fly without a sense of time, they would trace its eccentric behavior to that same first chromosome, the X. And when they mapped the mutant gene, they would locate it right next to Morgan's starting point, less than one map unit away from *white*.

# CHAPTER THREE

---

# What Is Life?

*With all his amateurish fumbling, Martin had one charac-
teristic without which there can be no science: a wide-
ranging, sniffing, snuffling, undignified, unself-dramatizing
curiosity, and it drove him on.*

—SINCLAIR LEWIS,
*Arrowsmith*

MORGAN HAD ENTERED the field as a critic, a gadfly, and he was never
as comfortable with the theory of the gene as those who came after him.
Unlike Sturtevant, Calvin Bridges, Curt Stern, and many of his other
gifted students, Morgan did not have a mathematical mind. He was a
born naturalist. He loved to work with starfish, sea anemones, and
pigeons, even though bottles of fruit flies eventually crowded them out of
his laboratory. In some ways he resembled the German physicist Max
Planck, who turned physics upside down in 1900, the same year that
biology rediscovered Mendel. Planck described a beam of light as a
stream of bits, packets, or quanta of energy, much as Morgan described
the transmission of life as a stream of bits. Planck's quantum theory was
so hard for a classical physicist to absorb that Planck himself spent years
fighting it. "He was a revolutionary against his own will," said one of his
students, James Franck. "He finally came to the conclusion, 'It doesn't
help. We have to live with quantum theory. And believe me, it will
expand. It will not be only in optics. It will go in all fields. We have to live
with it.' "

Morgan's discovery transformed biology as much as Planck's transformed physics, and Morgan was sometimes almost as ambivalent about
his revolution. In the war between "bug hunters" and "worm slicers," as

33

the two camps sometimes called themselves, the old-fashioned outdoor naturalists and the newfangled indoor experimentalists, Morgan was an outdoor man who brought biology indoors. He was a "squishy" who fought for the "crunchies." For years he struggled to keep up with Sturtevant, Bridges, and Stern as they crossed mutants, crunched numbers, and mapped the first, second, third, and fourth fly chromosomes. Morgan did manage to make contributions to the effort, but only through hard work and increasingly strained powers of intuition, as one of his students, Curt Stern, has written. "I remember the 'awe-full' moment when Bridges explained to him a particularly intricate new result and the initiator of it all left the room, shaking his head and saying 'too much for me!' "

Morgan also had trouble seeing the connection between their discovery and Darwin's. His students tried to explain it to him. What they were doing in the Fly Room is just what natural selection does in the wild: choosing and selecting tiny mutations of the kind that give a fly a red or white eye, a straight or forked bristle, a long or short wing. Over many generations this natural selection of small changes could produce two separate species of flies; and with more generations it could produce bigger and bigger branchings in the tree of life. Some of the greatest biologists of the twentieth century, including R. A. Fisher, Sewall Wright, J.B.S. Haldane, Ernst Mayr, G. Ledyard Stebbins, and Theodosius Dobzhansky (another of Morgan's students), would eventually unify Darwin's and Morgan's theories, bringing together outdoor and indoor nature, visible and invisible life in the synthesis. But in the Fly Room, Sturtevant used to explain Darwin's theory to Morgan over and over again. "You had to keep working on it," Sturtevant once told Garland Allen, T. H. Morgan's biographer, in an unpublished interview. "He wouldn't stay convinced about that. You had to keep at it. You had a job to do over again every once in a while."

It is a mark of Morgan's courage that he eventually moved his Fly Room to Caltech, which was, then as now, one of the world's greatest research centers for chemistry and physics. By shifting his operations there in 1928, at the age of sixty-two, Morgan hoped to help unite biology with chemistry, physics, and mathematics. Morgan was not grounded in those subjects himself, but like a general who sees where the war is going he wanted troops for the new front. He did have something of the general in him, by nature or nurture. His father, Charlton Hunt Morgan, had fought under General John Hunt Morgan (Charlton's brother, Thomas's uncle) in a legendary band of rebel daredevils known as Morgan's Raiders. Some of his more distant kin included J. Pierpont Morgan, the robber baron, and Francis Scott Key, the author of "The Star-Spangled Banner."

In Pasadena, Morgan and his raiders (including Sturtevant and Bridges, who never left the man they called The Boss) explored and promoted the theory of the gene; and when Morgan was awarded the Nobel Prize for the work, he shared the prize money with Sturtevant and Bridges. But even in his Nobel address in Stockholm in 1934, Morgan expressed some doubts about their discovery. "What are genes?" he asked. "Now that we locate them in the chromosomes are we justified in regarding them as material units; as chemical bodies?" Geneticists, he said, could put that question to one side, temporarily. They could work with genes as mathematical points on abstract maps. This was the kind of work that Morgan and his raiders did, and it would later become known as pure genetics, classical genetics; it was work without molecules, as opposed to the work that Benzer and his circle made possible. When Morgan looked at the points on the maps he still wondered "whether they are real or purely fictitious."


BENZER TOOK his first look through a microscope in 1934, the same year that Morgan asked, "What are genes?" Morgan was in Stockholm, receiving his prize. Benzer was in Bensonhurst, Brooklyn, and the microscope was a bar mitzvah present. Benzer carried it down to the basement, where he had built a laboratory. There he performed what seemed to him the obvious first experiment and stared down through his microscope at hundreds of thousands of long, dark, thrashing tadpoles with tiny heads: sperm.

No one else in the Benzer family cared for science. His parents came from a shtetl west of Warsaw, and they worked in the needle trades. His father would come home from the Garment District of Manhattan with bundles of clothes, which his mother would finish on her sewing machine late at night. Sometimes they asked Seymour to ride the subway to deliver their bundles. But Seymour was the only boy in the family—he was the Benzer prince, "the egg with two yellows," to use an Old World expression—and on most afternoons and evenings his parents and his three sisters left him free to play stickball on Sixty-eighth Street or to pursue his researches in the basement. After his bar mitzvah he had nothing more to do with Hebrew School. On High Holidays, Seymour would go with his father to the synagogue, because it was a shame for a

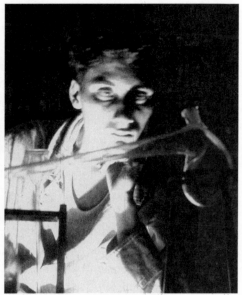

*Seymour Benzer's first laboratory, his family basement in Bensonhurst, Brooklyn. Here Benzer mixed potions, deconstructed flies, and struck mad-scientist poses. He also read* Arrowsmith, *the novel that later helped lead him to the study of the gene. He made these self-portraits, with his camera on a timer, in the mid-1930s.*

father not to have his son beside him on Rosh Hashanah and Yom Kippur. Even then he would smuggle in something to hold over the prayer book. While the rest of the congregation chanted and his father looked the other way, Seymour read Stern and Gerlach's *The Principles of Atomic Physics.*

Through the microscope he inspected blood, sweat, tears, spit, tongue gook, gutter water, and bee stings. Over and over again he deconstructed flies—house flies. His favorite book as a teenager was *Arrowsmith,* by Sinclair Lewis. It was almost a prayer book, because it showed science as an adventure, a romance, and a pure faith, a way to live a life. The hero of the novel, Martin Arrowsmith, is born in a small town in the Midwest. As a freshman at the University of Winnemac, he hears rumors about a mysterious German biologist on campus, a man named Max Gottlieb, who studies bacteria. Late one night, after a party, Arrowsmith wanders over to the medical building, stares up at the tall turrets of the Main Medical Building, and sees a single light. Even as he looks, the light goes out, and soon a man comes stooping along the path toward him, an old, gaunt man with his hands clasped behind his back, muttering to himself, and passes him on the path: Gottlieb. "He had worn the threadbare top-coat of a poor professor, yet Martin remembered him as wrapped in a black velvet cape with a silver star arrogant on his breast."

In the novel, Arrowsmith makes himself a disciple of Max Gottlieb, a man of "tyrannical honesty" who paces night after night in his laboratory and who tells his students, "I make many mistakes. But one thing I keep always pure: the religion of a scientist." From Gottlieb, Arrowsmith learns to scorn the kind of careerist in medicine who thinks only about the practice and the fee; or the kind of plodding scientist who "never ventured on original experiments which, leading him into a confused land of wondering, might bring him to glory or disaster." Arrowsmith ventures; he meets and marries a wonderful nurse named Leora; he finds both glory and disaster.

As soon as Benzer finished reading *Arrowsmith,* he bought the finest-pointed fountain pen and the blackest ink he could find and began to imitate Max Gottlieb's handwriting, just as Sinclair Lewis describes it in the novel, "that dead-black spider-web script."

When he graduated from high school at the age of fifteen, no one in his family had ever gone beyond the twelfth grade, and the Great Depression was dragging down his father's business. But Seymour was the egg with two yellows, and he entered Brooklyn College on a Regents Scholarship. There, in his freshman year, he met a twenty-one-year-old nurse named Dora, nicknamed Dotty. Dotty took night duty at the hospital so that they could sneak time with each other while the patients slept, just like Martin Arrowsmith and his Leora.


NOT MANY PEOPLE have both a feeling for physics and a feeling for the study of life. In the original Fly Room, the raider with the strongest interest in both these sciences was a short, high-strung, nervous, visionary young man named Hermann J. Muller. From the beginning, Muller was eager to prove that genes are solid objects and to find out what they are made of. This interest helped to set him apart from Morgan's other raiders, as Sturtevant later told the historian Garland Allen: "The rest of us would have gone along without, and thought, 'Well, maybe our great-grandchildren will know a little something about it'—this was kinda the attitude."

It was Muller who eventually found a way to accelerate the flies' mutation rates, the project that had frustrated Morgan in the years before the arrival of the white-eyed fly. Long after Muller graduated from Morgan's Fly Room to a Fly Room of his own, he managed to induce mutations in *Drosophila*. He did it by bombarding the flies with radiation, much the way atomic physicists were now inducing transmutations in atoms. Muller zapped flies with such powerful doses of X rays that he increased their mutation rates by about 15,000 percent. The X rays killed billions of sperm cells in the flies, but of the sperm cells that survived, almost half carried new mutations. Muller's flies still looked and acted the same after their radiation treatments, but their children looked different in the wings (*expanded, mussed, splotched*) or the hairs (*ruffled*), or the body (*stumpy, cloven thorax*). Some of these changes Muller noticed only when he knocked out the flies with ether and inspected them through the microscope. But many changes he could see with the naked eye. One of the mutants was a white-eyed fly.

In Germany, a young atomic physicist named Max Delbrück heard about these "raying" experiments and decided that the transmutation of genes might be even more interesting than the transmutation of atoms. He left Germany and joined Thomas Hunt Morgan at Caltech, where Morgan's experiment in the unification of the natural sciences was still just getting going. At Caltech, chemists and physicists felt a wide gulf between their world and the world of the biologists, and vice versa. When Delbrück found his way to the Fly Room, Sturtevant told him to start by clearing up certain small points of confusion in the map of the fly's fourth chromosome. Sturtevant handed Delbrück a stack of reprints, and Delbrück took them to a room across the hall. There he pored through them, more and more disconsolately. To him they were much harder than atomic physics: "Forbidding-looking papers, every genotype was about a mile long, terrible, and I just didn't get any grasp of it."

Bridges tried to help. A genotype is a description of the genes of a living thing: the lines of genetic code that it has inherited. A phenotype is the expression of the genes: the living thing itself, the way it looks or behaves. They had named their first gene *white* after the white-eyed fly. But the *white* gene comes in more than one form. Most flies carry two copies of the normal form of the *white* gene, and they have red eyes. Some flies carry one copy of the normal form and one copy of the mutant form, and they have red eyes, too, because the normal form is dominant, even though they have one mutant form of the gene in their genotype. Here and there a fly carries only the mutant form of the gene, and that fly has the signature white eyes. In spite of these complexities of genotype and phenotype, Morgan's Raiders called the gene *white,* and they also called a white-eyed fly *white.*

Now, every fly in the Fly Room carries not one but many mutant genes. So does every fly outside the Fly Room (known in the jargon as "wild-type") and so does every human being (wild or not). Most of these mutant genes are not expressed. Drosophilists often breed flies to make double mutants, triple mutants, and more. A fly's genotype, scrawled on the label of a fly bottle, may begin *f; cn bw; TM2/tra* . . . Any young Lord of the Flies can read that formula at a glance. It means the fly in the bottle carries the mutation *forked* on its X chromosome. The word *forked* is short for *forked bristles.* The fly carries *cinnabar* and *brown* on its second chromosome. Although *cinnabar* and *brown* both refer to eye color, when both mutations come together in a single fly they tend to produce white eyes. The fly also carries *tra,* or *transformer,* on its third chromosome. If a fly with two X chromosomes has *tra* on its third chromosome it grows up with a male's inky-black abdomen, a male's penis, a male's sperm, and it mates with females, even though having two X chromosomes makes it, genetically speaking, a female fly. *TM2* is a long story. . . .

Delbrück liked Sturtevant and Bridges, but he could not absorb all this lore and jargon. He had grown up in Berlin on the same block as Max Planck. As a student he had rubbed shoulders with Niels Bohr, Werner Heisenberg, Erwin Schrödinger, Albert Einstein. Now he missed the deep, clean simplicity of Planck's $E = nh\nu$ and Einstein's $E = mc^2$. He felt he had to get out of the Fly Room or die. Eventually he found a microbiologist down in the basement who was experimenting with bacteria, and with viruses that kill bacteria. The bacteria were *Escherichia coli,* the common bacteria of the human gut. The viruses were bacteriophage (from the Greek *phagein,* "to devour"); phage for short. *E. coli* was a well-known organism. Bacteriophage was well known too, partly through fiction, as Arrowsmith's path to glory and disaster.

In bacteriophage, Delbrück saw a way to escape from the Fly Room. With phage and *E. coli* he could reduce the phenomena of inheritance to the kind of deep, clean, simple problem that he loved in physics. He allowed *E. coli* bacteria to expand like a living carpet across the bottom of a petri dish. Then he scattered some predatory particles of bacteriophage into the dish, and within a few hours the phage ate holes in the carpet. A bacterium is too small to see with the naked eye, and a single virus particle is too small to see even with the most powerful light microscope. But Delbrück could distinguish different strains of phage just by glancing at the holes they ate in the carpets: some strains ate big, shaggy holes, some ate neat, tiny ones. In other words, as Delbrück once put it, strains of viruses make themselves known by their behavior, the way "a small boy announces his presence when a piece of cake disappears." And these differences in behavior are in the genes.

During the World War II years, Delbrück stayed on in the United States. (He and his family were ardent anti-Nazis; it would have been dangerous to go back to Berlin even if he had wanted to.) He and a small but expanding circle of friends, including another resident enemy alien, the Italian biologist Salvador Luria, carried out a series of phage experiments—elegantly simple experiments, in the style of physicists—trying to find the particles of heredity and figure out how they work. They were going now where Morgan had dreamed his students would go, investigating the behavior of copulating chromosomes at a level beyond reach of the microscope. ("They must come together with extraordinary precision," Morgan had written, "which implies probably that we are dealing with events of a molecular order. We can go no further until physics has furnished us with a key to unlock these extraordinary events.") A key discovery came in 1944, when a microbiologist's study of bacteria hinted that the crucial substance might be deoxyribonucleic acid, or DNA. But not many biologists noticed that study. Morgan himself was now in his last years—he died in 1945—and he did not dream that the answer to the question was close. "He just didn't feel as though there was any toehold," Sturtevant told Garland Allen, the historian, in their unpublished interview. "I remember one of the last times I ever talked to Morgan—not very long before his death. He said he hadn't been keeping up with things, but *this* is the thing he'd *always* wanted to know." What are genes? What passes between the chromosomes?

"This was the atmosphere," Sturtevant summed up, speaking for the vast majority of biologists who were still in the dark in 1945. " 'Boy, if we could only get going on it—but how?' Nobody had any ideas."

SEYMOUR MARRIED DOTTY in January of 1942, just as the U.S. swung into the war, and the two of them hopped on a train for Purdue University in Indiana, where he enrolled as a graduate student in the physics department. Almost immediately he was recruited for a secret wartime project. Physicists in a far-flung Allied effort that rivaled the Manhattan Project in both scale and military significance were trying to build a new, improved radar. At that time, a radar set was built around a device called a silicon crystal rectifier, which is a sort of one-way turnstile for electric-

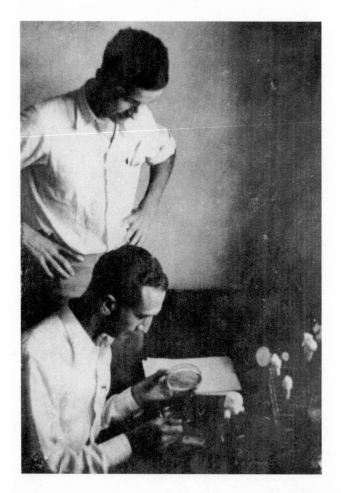

*The young atomic physicist Max Delbrück advanced the study of the gene through elegant experiments with viruses and bacteria in petri dishes. Here Delbrück looks over the shoulder of his first partner in these experiments, Salvador Luria, at Cold Spring Harbor Laboratory in 1941, the year the two men began working together. Their meeting was the beginning of molecular biology and would lead to a joint Nobel Prize.*

ity: in a rectifier (unlike an ordinary copper wire) electric current can flow in only one direction. These rectifiers were unreliable, and the leader of the Purdue team, a Viennese physicist named Karl Lark-Horovitz, was trying to replace the silicon with germanium. Most of the physicists on the team were young and inexperienced, but then, so was the science of electronics. Physicists did know that germanium and silicon conduct electricity better than wood and worse than copper, which is why they are called semiconductors. But for some aspects of semicon-

*As a student of physics and chemistry at Brooklyn College, Benzer peers at the spectrum through a spectroscope. Biology needed help from chemistry and physics, but not many people in the late 1930s had a feeling for all three sciences. Today much of the science of molecular biology traces back to Benzer and a few other physicists-turned-biologists.*

ductor behavior, the physicists were still in the dark—which is where Benzer, from the beginning, has done his best work. He made some of the basic discoveries that led to the construction of a stable germanium rectifier. Among other things, he discovered a germanium crystal that could withstand very high voltages.

Earlier in the twentieth century, physicists had thought of the electron as an otherworldly curiosity. Young physicists at the annual Christmas dinner of the Cavendish Laboratory in Cambridge used to shout a toast: "The electron! May it never be of any earthly use to anybody." But of course even pure, romantic, otherworldly research—Arrowsmith's kind of research—can change the world. During the war, Purdue's germanium crystal recitifers made their way to Bell Laboratories in New Jersey. There a team of physicists, using Purdue's germanium, took the semiconductor a crucial step further after the war was over and invented the transistor. Transistors are the central element of radios, televisions, computers—all things electronic. They started a new industrial revolution; electronics is now the biggest industry in the United States. So Benzer had helped put the electron to earthly use, at the age of twenty-three.

In 1946 a friend of his in the secret laboratory passed him a book by

*Seymour and Dotty Benzer got married in New York City in 1942 and hopped on a train to Purdue University, in Indiana. There Benzer would help start his first revolution—electronics.*

the German quantum physicist Erwin Schrödinger, *What Is Life?* It was a book about the gene problem. Schrödinger had written it in Dublin, where he had gone to escape the war. For a physicist to think about biology was itself a kind of escape from the war.

In *What Is Life?* Schrödinger tried to connect the world of atomic physics with the world of genetics. He suggested that the gene might turn out to be a novel kind of crystal, an aperiodic crystal, in which the message is locked into the crystal lattice like a series of letters, and those letters might carry the secret of life. "We seem to arrive at the ridiculous conclusion that the clue to the understanding of life is that it is based on a pure mechanism," Schrödinger wrote. That is, the secret of life is nothing more than a kind of clockwork. "But please, do not accuse me of calling the chromosome fibres just the cogs of the organic machine," he wrote. Any cog that does what these cogs can do is "not of coarse human make"; it must be "the finest masterpiece ever achieved along the lines of the Lord's quantum mechanics."

To Benzer, Schrödinger was thrilling on the theme of the hereditary substance, whose permanence, he noted, is "almost absolute. For we must not forget that what is passed on by the parent to the child is not just this or that peculiarity, a hooked nose, short fingers, a tendency to rheumatism, hemophilia, dichromasy, etc." What is passed on is not only three-dimensional but four-dimensional: a whole human being moving through time.

*What Is Life?* did not offer any hard news about the nature of the gene. The book's chief reference was some obscure work that Max Delbrück had done before he left Germany on the idea that a mutation is like a quantum jump. Schrödinger made that idea his book's centerpiece: he called it "Delbrück's Model." Actually, that work of Delbrück's was already years out of date; because of the war and the walls that still existed then between physics and biology, Schrödinger knew nothing about Delbrück's work with phage. Still, *What Is Life?* had an enormous effect on a whole generation of young scientists because it made the gene problem sound like *the problem to solve.* At the Admiralty Headquarters in London, in a windowless pile known as the Citadel, the young physicist Francis Crick, who had spent the war working on mines, read *What Is Life?* and decided to become a biologist. At the University of Chicago, an undergraduate named James Dewey Watson picked up the book. He had been studying birds, "but," he later wrote, "from the moment I read Schrödinger's *What Is Life?* I became polarized toward finding out the secret of the gene."

At Purdue, reading *What Is Life?* made Benzer wonder if his own doped crystals of germanium might somehow be related to the mysterious crystals of inheritance. Benzer could not help lingering a little sentimentally over the name Max Delbrück, which has the same romantic aura of Germanic genius in *What Is Life?* as the name Max Gottlieb has in *Arrowsmith.* Through the grapevine, Benzer learned that Delbrück was now working at Caltech and that he was working on viruses and bacteria, like Max Gottlieb. Benzer had already found his Leora—this sounded like his Max. At a meeting of the American Physical Society in Bloomington, Indiana, Benzer was invited to dinner at the home of a biologist by the name of Salvador Luria. Benzer asked Luria if he happened to know anybody who worked on viruses.

"Well, yes, *I* work on viruses," said Luria.

"Tell me," Benzer said, "did you ever hear of Delbrück?"

Luria went to a drawer and pulled out a snapshot of Delbrück. He and Delbrück had been working together since 1940; Luria was Delbrück's first serious collaborator in bacteriophage. Benzer could not have been more impressed. Before the evening was over, Luria was urging him to take a summer course in bacteriophage at Cold Spring Harbor, a course that Delbrück had established a few years before in order to do just what Morgan had been trying to do—to draw more hard scientists, and especially more physicists, into the biological fold.

Benzer did take the phage course that summer. By now Delbrück, Luria, and their friends had begun to speak in a jargon of their own to describe a change in the fate of a cell: induction, transformation, cell determination, cell commitment. Within one day, Benzer says, he became instantly induced, transformed, determined, and committed to be a biologist. When they got back to Purdue, he and Dotty drove out of Lafayette, parked out in the Indiana flatlands, and talked things over. Not all children of the Great Depression would have considered the choice that they were looking at. They had a one-year-old baby girl. Seymour's family had been poor, and Dotty's had been poorer. She had worn one skirt all through high school, very shiny in the back.

"People in my group at Purdue thought I was nuts," Benzer likes to say now. "Here it was, the semiconductor thing was booming, Lark-Horovitz and I had gotten six patents on the work we had done with the semiconductor." The other veterans of the secret lab at Purdue were already planning to form electronics companies and get rich. "Are you crazy!" they told him. "You can ride in on the tide!"

But ever since his Arrowsmith years, Benzer had believed that happiness is the pursuit of curiosity and that a fall from pure science is a fall from grace. "I'm interested in biology," he said simply. And Lark-Horovitz, who must have seen that his golden boy was already gone, gave him his blessing.

CHAPTER FOUR

# The Finger of the Angel

*I study myself more than any other subject. That is my*
*metaphysics; that is my physics.*

—MICHEL DE MONTAIGNE,
*"Of Experience"*

BENZER KEPT HIS BASE in the physics department at Purdue, but he
began living like a gypsy. In the late 1940s and early 1950s he spent a year
at Oak Ridge National Laboratories in Tennessee; two years in Delbrück's laboratory at Caltech; a summer at the laboratory of Cornelius
van Niel at Pacific Grove; and a year in André Lwoff's laboratory at the
Institut Pasteur in Paris. In all these places Benzer helped establish the
style of the phage group by doing simple and elegant experiments
("pretty and witty," in the words of Horace Freeland Judson, the historian
of molecular biology). Like Arrowsmith's mentor Max Gottlieb, Benzer,
from the beginning, was a scientist's scientist. He kept a low profile, and
he worked mostly in the middle of the night.

It was Delbrück who set the tone for the phage group. Even in that
crowd, Delbrück was intimidatingly intelligent. He was also young,
quick, fit, and mordantly funny, with a young and beautiful wife. Those
who followed Delbrück found him so charismatic that they let him treat
them the way a Zen master treats his disciples; he threw them into the
mud again and again to help them achieve enlightenment. Delbrück
always chose a front-center seat at seminars. That way when he sprang
up in the middle of a talk he would block the slide projector and force
half his row to let him pass as he struggled toward the door, denouncing
the lecture as he went. Everyone who visited Delbrück's group at Cal-

tech was obliged to give one of these seminars, and Delbrück always pronounced the same verdict: "This was the worst seminar I ever heard."

The great breakthrough in the gene problem arrived unannounced one morning in April 1953, in a tower room of the Cavendish Laboratory in Cambridge—already a legendary place in physics, where J. J. Thomson had discovered the electron and Ernest Lord Rutherford had split the atom. When James Watson and Francis Crick put together their model of the double helix, they accomplished in a few minutes what Morgan's Raiders had been trying to do for decades. Physics, chemistry, and biology came together in one beautiful spiraling molecule, the staircase of DNA, now an icon of twentieth-century science together with the fly bottle and the mushroom cloud. Crick was still in his thirties, and Watson was one month away from his twenty-fifth birthday. In snapshots from that time he looks like a boy, though he has a lean and hungry look. Watson used to run around Cold Spring Harbor in short pants and with each sneaker trailing the double strands of its laces. Long hair, shorts, and unlaced sneakers were his signature, Benzer remembers, his way of shocking the bourgeois. ("He used to really infuriate André Lwoff by showing up in meetings in France with his shoelaces deliberately untied.")

Watson and Crick saw at a glance that they had not only solved the physical structure of the gene; they had found the way it carries the secret code. The spiral staircase of DNA holds the secret in the treads, the small molecular crosspieces that are known as bases or nucleotides, and these bases come in four chemical varieties. Schrödinger's *What Is Life?* had pointed out that even a small number of signs can make an alphabet. In Morse code there are just two different signs, dot and dash. Schrödinger had predicted that the code of life might turn out to have only a few signs as well. In fact it has four, the four treads of the twisted stair: adenine, cytosine, thymine, and guanine, or A, C, T, and G. The spiral staircase can hold any sequence of bases, any permutation of letters, A, C, T, G, A, G, C, A, and so on, millions of letters in a single strand of DNA, three billion letters coiled and supercoiled in the nucleus of every human cell.

CRICK SAYS THAT WHILE they were writing a report for *Nature*, "A Structure for Deoxyribose Nucleic Acid," Watson "suffered from periodic fears that the structure might be wrong and that he had made an ass

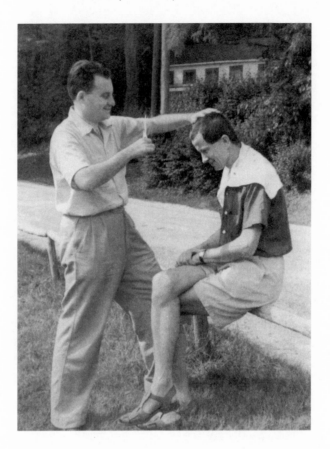

*Cold Spring Harbor Laboratory, early 1950s: revolution on five dollars a day. Max Delbrück and many of the first molecular biologists who gathered around him were young and poor, and they created their new science in bohemian high spirits. "Max had a tradition of trading haircuts with people," Seymour Benzer says (he's doing the cutting here). "He made a deal with me on this occasion that each of us would cut the other's hair, but the one who was cut first was not allowed to look in the mirror beforehand."*

of himself." And for some years afterward neither Watson nor Crick could rest easy. It was one thing to say that they had found the code of life, another to prove it. On the maps of classical geneticists, *white, yellow,* and *miniature* were dots, abstractions. They did not look like long twisted-chain molecules; they looked the way planets had looked to astronomers before telescopes or the way atoms had looked to physicists at the turn of the century, indivisible and indestructible points.

*A center of the revolution: Max Delbrück's phage laboratory at Caltech. Delbruck sits by the window, Gunther Stent in the middle of the huddle. The two men would later be among the first to turn from the study of the gene to the study of genes and behavior.*

So after the eureka of Watson and Crick, one of the challenges for the new science (which did not yet call itself molecular biology) was to connect these classical maps of the gene with the new model of the double helix. It was Benzer who thought of a way to do it. Not long after Watson and Crick announced their discovery, Benzer hit on a plan that might unite the old revolution and the new revolution: classical genetics and molecular biology.

Benzer's starting point was, as usual, "pretty and witty." He decided to go back to the event that had opened the science of genetics, Sturtevant's big night. When two chromosomes line up and exchange bits of genetic material, Benzer reasoned, many genes must cross over together en masse. Since each gene on the map looks like an indivisible point,

*Benzer (right), in the summer of 1953 at Cold Spring Harbor, planning the rII experiment.*

classical geneticists had always assumed that during crossing-over the chromosome always breaks between genes, just as when the blades of a scissors cut through paper they always pass between atoms. But Benzer knew that if Watson and Crick were right about the double helix, then each gene is not a mathematical point with open space around it. Instead, a gene must be a long, continuous, twisted thread, a string of rungs or nucleotides. If it is a molecular construction consisting of millions upon millions of atoms linked together, there is no particular reason why a break should not fall within a gene as well as between two genes, just as if one rips a piece of newspaper at random the tear will go through words as easily as it goes between them.

A few of Morgan's Raiders had speculated the same way and tried to explore the idea. One of Morgan's students' students, Guido Pontecorvo, had written a brilliant paper on the subject; a few others had managed to rip a fly gene once or twice, in heroically laborious experiments. Now, in the light of the double-helix model, in which a gene is made of the rungs in a long, twisted ladder, these speculations and experiments seemed more compelling. Any spaces between the genes must be made of rungs of the same material as the genes themselves. In this model, there is no

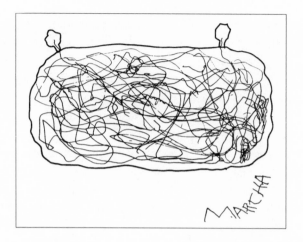

*By chance, two particles of virus (at top) have attached themselves to a single E. coli bacterium and injected their long strands of DNA. In 1953, the year of the discovery of the double helix, Benzer invented a way to use this mingling of viral DNA to map the interior of a gene. This illustration comes from one of the historic papers in which Benzer reported the results of the experiment. Benzer's caption: "The artist, Martha Jane Benzer, who graciously signed the drawing, was five years old at the time."*

obvious reason why during crossing-over a thread of DNA should not sometimes break right in the middle of a gene. If genes do sometimes break in the middle, and if Benzer could find one of those breaks, he thought he could join the old science of the gene with the new science that he and his friends were creating and lift them both to a dizzying new level.

Benzer's plan required him to arrange matings between strains of virus, just as Mendel had crossed peas and Morgan had crossed flies. Viruses do not have sex. But Benzer could get around that problem by infecting a plate of bacteria with two strains of phage at once. Here and there on the plate, two virus particles, one from each of the two strains, might come together in attacking a single bacterium. This event would later come to assume so much importance for Benzer that his younger daughter, Martha, at the age of five, would be moved to draw a picture of that rare event, the double infection.

Each virus has only a single chromosome. But inside a hapless twice-bit bacterium, the chromosome from one virus particle would twine together with the chromosome of the other. Then the two chromosomes would twist like copulating coral snakes, just like pairs of chromosomes in peas, flies, and human beings, and some of their genes would cross over.

By 1953, phage workers had already mapped much of the phage chromosome. On their maps, a mutation called *r* appeared as a mathematical point in a chromosome region called *rII*. The *r* stood for "rapid": *r* mutants devour bacteria fast. Benzer arrived at the idea for his now legendary experiment when he stumbled across a strain of defective *r* mutants—a strain of mutants that was, so to speak, off its feed—and he decided to focus on the *rII* region.

He would cross two separate strains of defective *r* mutants in a petri dish. In the classical view of genes and mutations, the two strains of mutants would have identical *rII* regions and would produce nothing but defective children. But if the Watson-Crick view was right, then the damage in each of two strains of *r* mutants might lie at two different points inside that region. By crossing two defective *r* mutants, he might be able to prove that. Suppose, for example, one parent carried genetic damage at one end of its *rII* region. Suppose the other parent carried damage at the opposite end of its *rII* region. And suppose that when the two chromosomes twined together, they happened to trade bits of the *rII* region. Then the mosaic chromosome they put together inside the bacterium might contain a healthy chunk of the *rII* gene from one parent and a healthy chunk of the gene from the other parent. Their children would be healthy.

So if Benzer crossed two defective *r* mutants and got one or more healthy *r* children, their arrival would prove that crossing-over can sometimes cut right through a gene, not just between genes. That would mean that genes, like atoms, are not indivisible points but solid objects that can be cut and dissected. If Benzer could in fact dissect a gene, he foresaw that he and his friends would soon be able to take his experiment much, much further.

Benzer realized all this one fine day in the fall of 1953 in his laboratory on the third floor of the physics building at Purdue University, where he was still (nominally) a professor of physics. The year before, in the course of a routine experiment with *r* mutants at the Institut Pasteur in Paris, he had stumbled across a defective *r* strain. Benzer remembers shrugging and throwing them out: "As Pasteur would say, 'My mind was not prepared.'" Now, at Purdue, while arranging a bacteriophage experiment for a classroom demonstration, he came across another defective *r* mutant. At first he thought he had made a mistake. "*Dummkopf*, do it again." He prepared a fresh carpet of bacteria and added more *r* mutants. But when he came back a little later, he saw that the *r* mutant still did not behave like a normal *r* mutant. Now his mind was prepared.

After much thought and a few summer-long conversations in Cold Spring Harbor with phage friends, Benzer wrote out a sketch of his plans for *rII*. Toward the end of the summer of 1954 he ran into Delbrück at a meeting in Amsterdam and showed him the sketch. By now Delbrück was the elder statesman of their revolution, just as Morgan had been the elder statesman of the old revolution, and Delbrück thought Benzer's *rII* manuscript was outrageous. The very idea that a gene might be split into pieces seemed to irritate Delbrück, Benzer says. "One of his typically succinct comments was 'Delusions of grandeur.'" Benzer still cherishes the comments that Delbrück scribbled on his manuscript: "You must have drunk a triple highball before writing this. This is going to be offensive to a lot of people that I respect."

Even assuming that Benzer's reasoning was correct, the chance of crossing two defective *r* parents and producing normal *r* children was astronomically low, on the order of one in a billion. At least, that was what his calculations suggested: he would have to breed enough virus to detect one-in-a-billion events reliably. But there would be more than enough particles of virus to do an experiment like that in a petri dish. "One can therefore perform in a test tube in twenty minutes," Benzer later wrote, "an experiment yielding a quantity of genetic data that would require if humans were used virtually the entire population of the earth." And Benzer saw all this and more in that first instant in his physics lab at Purdue, when he looked at the defective *r* mutant with a prepared mind. As Judson writes in *The Eighth Day of Creation,* "There was no way to see it except instantly."

In essence it was a very simple experiment, like all of Benzer's experiments, and almost from the moment he began he was splitting genes into pieces. He plunged into a whirlwind, like the young Martin Arrowsmith when he discovers phage: "Then his research wiped out everything else, made him forget Gottlieb and Leora . . . and confounded night and day in one insane flaming blur as he realized that he had something not unworthy of a Gottlieb, something at the mysterious source of life." In his petri dishes, genes and mutations finally ceased to be abstractions. The splitting of atoms by Rutherford had led to the atomic bomb, and the splitting of genes by Benzer would lead to the explosions of genetic mapping and genetic engineering that now dominate biology. For a few years his research made him forget everything else (except Dotty—they were uncommonly close). The excitement was particularly intense for lapsed physicists like Benzer and Crick, who had jumped from the flagship of the sciences for a small open boat in a wide sea. Crick asked Benzer to speak at the Kapitza Club in Cambridge, an exclusive club of physicists. (The discovery of the neutron was first announced there.) In the audience was Paul Dirac, one of the most powerful theoretical physicists of the century and also one of the quietest—much quieter than Benzer. Physicists visiting him at Cambridge were satisfied if they heard him say a single yes or no. "At least," they would tell one another, "I got a word out of Dirac."

Benzer opened his talk at the Kapitza Club by writing on a blackboard the date 1808, when John Dalton had published *A New System of Chemical Philosophy.* Next Benzer wrote the date 1913, when Bohr had published "On the Constitution of Atoms and Molecules." One hundred five years had passed between the first clear description of atoms as a possibility and the first clear description of atoms as a physical reality.

Then Benzer wrote on the board the date 1866, when Mendel had published his paper about peas, and the date 1953, when Watson and Crick had published their paper about the structure of the gene, the double helix of DNA. Only eighty-seven years had passed between the first clear description of genes as a possibility and the first clear description of genes as a physical reality.

Dirac looked at the blackboard and said four words: "Biology is catching up."


AFTER HE SPLIT the *rII* gene, Benzer spent a few manic years doing nothing but collecting *r* mutants and crossing them two by two. A friend and mentor of his in the phage world, Alfred Hershey, had once offered this definition of heaven: "To find one really good experiment and keep doing it over and over." Benzer felt he had found Hershey Heaven. In each mutant the *rII* region of the chromosome carried an error somewhere in the string of rungs of its DNA. He could use each error exactly the way Sturtevant had used his half-dozen mutations when he invented gene mapping. If two letters inside a gene are close together, their chance of being parted by crossing-over is small. If two letters are farther apart, their chance of being parted by crossing-over is correspondingly large. So whenever Benzer found a new mutant strain of *rII* in his petri dishes (these *r* mutants arise spontaneously and with some frequency in petri dishes, just as white-eyed flies arise spontaneously in fly bottles) Benzer could determine precisely where that particular copy of the *rII* gene was damaged. In other words, he could use the same method that Morgan's Raiders had used to map the locations of genes on chromosomes to map the relative positions of mutations inside the *rII* region. He was making the first detailed map of the interior of a gene. In the novel, when Arrowsmith discovers bacteriophage, he leaves his laboratory dawn after dawn, "eyes blood-glaring and set," and after a few weeks goes slightly mad with tension and exhaustion, "obsessed by the desire to spell backward all the words which snatched at him from signs." Benzer, driving home from his laboratory dawn after dawn on the long flat roads of Indiana, noticed his mind playing the same tricks: POTS. DEEPS TIMIL. TIMIL DEEPS. TIXE.

By the summer of 1956, he had mapped hundreds and hundreds of bits of the *rII* gene. He recorded them all on a mural that stretched farther and farther across his laboratory wall in the physics building. It was the world's first version of what would come to be called the sacred text, the code of codes. Biologists of a certain age can still remember the impression that Benzer made with his gene map at conferences, bearing it up on stage and unrolling it like a Torah scroll. If the single chromosome of a phage were stretched out in a straight line and magnified 150,000 times, it would be about ten meters long. At that magnification, the *rII* region would be about half a meter long. Benzer's scroll mapped the fine structure of that half-meter, with hundreds of different damage points inside.

To this day his old phage cronies from Cold Spring Harbor talk about Benzer and *rII* with awe:

"This is the atom breaker of biology."

"What he did in fine structure was epochal."

"He spent all summer at Cold Spring Harbor talking about the *rII* idea. I could have stolen it. I could have gone into my lab and done it myself. We didn't do that in those days."

Throughout the 1950s Benzer's scroll map got longer and longer. The gene was no longer a dot, a distant planet seen with the naked eye. The gene was the new territory of molecular biology. In 1959, when a geneticist put together a retrospective volume called *Classic Papers in Genetics*, he began his anthology with Mendel's peas, as a point of origin, and ended with Benzer's *rII*, as the point of origin for whatever would come next. So it proved. By the last years of the century, gene mapping would have grown into a project costing billions of dollars, the Human Genome Project, often called the Manhattan Project of biology. International teams of molecular geneticists would be racing one another to map every fly gene, every worm gene, and every last human gene at a cost of billions of dollars and at rates of more than one hundred million letters per year.

*"The atom-breaker of biology." From late 1953 through the early 1960s, Benzer worked on his map of the interior of a gene. He kept the map on a lengthening scroll. Here he poses wearily for a Purdue publicity picture with the scroll unrolled on his laboratory bench.*

But in the 1950s all this work was still obscure. It was remote from the shapes, colors, and visible wilderness that attracts most biologists to study life in the first place. Even in 1959, most biologists did not understand Benzer, any more than most biologists had understood Morgan back in 1911. He was mapping continents the rest of the world knew nothing about. In the summer of 1959, giving himself a break from his scroll, his years of "hard *rII*," as he put it, Benzer took a course in embryology at the Marine Biological Laboratory in Woods Hole, Massachusetts. During one evening lecture he was startled when the professor happened to mention the word "gene." Suddenly Benzer realized that he hadn't heard that word all summer, or the word "mutation."

"Yes, but what is a mutation?" one of the students asked.

"Oh, that's a very deep problem," said the professor. "We don't know anything about that."

"My God, what am I doing here?" Benzer thought to himself. "I'll go back to my genes and my mutations."

Eventually, to give people a better idea of what he was mapping, Benzer began collecting typographical errors from newspapers (which were now full of stories of the Cold War). Typos come in different categories. There are substitutions, places where one letter replaces another:

> . . . already the doomsday warnings are arriving, the foreboding accounts of a Russian horde that will come sweeping out of the East like Attila and his Nuns.

> —*Boston Globe*

And deletions:

> "I can speak just as good nglish as you," Gorbulove corrected in a merry voice.

> —*Seattle Times*

Insertions:

> "I have no fears that Mr. Khrushchev can contaminate the American people," he said. "We can take in stride the best brain-washington he can offer."

> —*Hartford Courant*

Inversions:

> He charged the bus door opened into a snowbank, causing him to slip as he stepped out and fall beneath the bus, which ran over him.

> —*St. Paul Pioneer Press*

And nonsense:

> Tomorrow: "Give Baby Time to Learn to Swallow Solid Food." etaoin-oshrdlucmfwypvbgkq

> —*Youngstown* (Ohio) *Vindicator*

Benzer was finding and mapping many of these sorts of mistakes in the *rII* gene: insertions, deletions, and nonsense. Mutations come down to nothing more than typos. The single chromosome of the phage virus contains about 200,000 letters of genetic code, about as many letters as there are in several pages of newspaper, so even in a single virus there is plenty of room for typos. And mutations that affect a fly, a mouse, or a human being in their repertoires of fundamental behavior will come down to typos too. In a sense, *rII* itself is a behavior gene, since damage there affects the behavior of the virus. Damage at any one of the thousand points in Benzer's map produced an identical change in the behavior of the virus. That is, a typo at any one of those thousand points in the map would wreck the *rII* gene. There is an old saying, "Each finger can suffer." In the genome, each gene can suffer, and each letter in a gene can suffer too.

In the summer of 1960, in the basement of Caltech's Church Hall, the physicist Richard Feynman took up what was known in those days as "Benzer mapping." Feynman loved Benzer's tricks for finding the single rare phage particle he was looking for in a dish of bacteria. He told friends it was like "finding one man in China with elephant ears, purple spots, and no left leg." And soon afterward in the Cavendish Laboratory, where Watson and Crick had put together the double helix, Crick and Sydney Brenner used *rII* mutants and Benzer mapping to help crack the genetic code. Crick and Brenner knew that they were looking at a four-letter code, A, T, C, and G. In an ingenious series of *rII* experiments, they proved that the code is a triplet. That is, the words are written in groups of three letters: CAT, TGA, AGT. Not long afterward, Benzer went to a meeting in India. Wandering in the street markets looking for exotica—he had acquired a taste for strange foods as well as strange hours—Benzer saw a soothsayer with a bird. Passersby would ask the soothsayer a question. Then the man would ask the bird. The bird would go into the cage, peck among the scraps of paper on the floor, and bring out an answer. Benzer asked, "Is the genetic code universal?" The bird gave the answer "The news from home is good."

In Paris, in the attic of the Institut Pasteur, Jacques Monod and François Jacob explored some of the implications of the new view of the gene. Each of us starts out as a single cell, and each of us ends up a collection of cells of many different kinds. Yet each of our cells still contains the same set of genes as the first. In a sense each of our cells knows everything but expresses only a small part of what it knows.

There is a story in Jewish legend that each baby arrives knowing everything, which accounts for the infinite wisdom we see in the faces of newborns. As the baby comes out into the world an angel places a finger just above its mouth to keep it from expressing all this wisdom, which accounts for the philtrum, the crease above the upper lip. Somehow something like the finger of the angel must touch our DNA and keep most of it from being expressed in each of our cells. Some genes turn on only in a liver cell and others only in a brain cell. Much later, in the 1990s, when molecular biologists began mapping all of the genes in the human body, they would discover several thousand genes that switch on only in the neurons of the brain—twice as many genes are expressed in the brain than are expressed anywhere else in the body. But no cell ever reads all of the words on the scroll. So every living thing and every last cell in our own bodies can say, with the preacher of Ecclesiastes, "When I travelled, I saw many things; and I understand more than I can express."

In their rabbit warren of laboratories at the Institut Pasteur, Jacob and Monod discovered the finger of the angel. They identified what they called "repressors" that float through the cell's nucleus and, by touching the double helix here and there—attaching themselves to strategic places all along its coils—silence most of the genes in most of our cells most of the time, so that only a small part of each double helix is expressed at any moment. The few genes that the cell does need are expressed; the rest only stand and wait. Today molecular biologists can actually watch this happen. Using an instrument called an atomic force microscope, they can watch enzymes sliding down strands of DNA and they can see the strands of DNA unrolling, very much like Torah scrolls, to begin reading or cease reading the portion of the scroll that is appropriate for that moment.

Jacob and Monod knew that these angelic floating proteins were a first glimpse of the connection between genes and behavior. They were looking at the beginnings of the senses, the tools with which a living thing picks up changes in the environment around it and uses the information to shape its behavior. Everything depends on such small felicitous moments of recognition: on compound meeting compound, shape meeting shape, profile recognizing profile. The shapes of these floating proteins allow a cell to recognize new chemicals entering the cell and to read just the portion of DNA that it needs at each moment in order to respond to each small event in its vicinity. Ezra Pound wrote a poem after spotting friends in the Paris Métro:

> *The apparition of these faces in the crowd*
> *Petals on a wet black bough.*

The fingers of the angels in every one of our cells are engaged in these small moments and shocks of recognition, not just when we are born but every moment of every day in every one of our cells.

Toward the end of the century a molecular geneticist working with a flock of sheep in Scotland would discover a way to give a cell a little shock of electricity and make the angels, just for a moment, snatch back their fingertips. His work would suggest that any cell—even a cell scraped from the inside of a ewe's udder or from a human cheek—can be made to express every bit it knows and grow into a lamb or a human baby, philtrum and all.

Even in the 1950s, the first molecular geneticists knew they were moving into strange new territory, and Benzer's taste for strange food and strange hours seemed of a piece with it. Benzer worked with Crick in his tower room at the Cavendish, with Jacob and Monod in their mansard rooms at the Institut Pasteur, with Delbrück in a basement at Caltech, and everywhere he went they told stories about his behavior. When Benzer was at the Pasteur, he shared a laboratory with Jacob. Jacob remembers Benzer in his memoirs: "Every day, at lunch, he brought some unusual dish—cow's udder, bull's testicles, crocodile tail, filet of snake—which he had unearthed on the other side of Paris and which he simmered on his Bunsen burner."

Benzer ate like that at home too: caterpillars, duck's feet, horsemeat, live snails. His little girl Barbie woke up one morning in Paris with her eyes swollen shut, and he took her to a doctor, who asked, "Has she eaten anything unusual lately?" Benzer was too mortified to be truthful.

"During the first months, there were few exchanges between us," Jacob writes, describing his labmate in his memoirs. "We did not keep the same hours. I arrived at nine in the morning; he, around one in the afternoon. As he came in, he would throw out a resounding 'Hi!' and then, after lunch, immerse himself in the inspection of his cultures. During the afternoon, he would belch once or twice. Around seven o'clock in the evening, I would bid him good-night and leave him to his nocturnal experiments."

CHAPTER FIVE

---

# A New Study, and a Dark Corner

*Psychology was to him a new study, and a dark corner of education.*

—HENRY ADAMS

WHEN BENZER was a physicist, he used to marvel at the power of mathematics to predict large portions of the behavior of the universe, from the fall of an apple to the arc of a rocket, from the quantum jump of an electron to the shining of the sun. No one could explain the connection between a brief formula and an apple, a rocket, an electron, or a star. No one could understand what one physicist called "the unreasonable effectiveness of mathematics." A mathematical physicist who helped invent the atomic bomb once wrote, "It is still an unending source of surprise for me to see how a few scribbles on a blackboard or on a sheet of paper could change the course of human affairs."

After the discovery of DNA, the whole world marveled at the unreasonable effectiveness of molecules. It was easy to see that the science that Benzer, Watson, Crick, Sydney Brenner, Gunther Stent, and a few others helped to establish in the middle of the twentieth century might someday change the course of human affairs even more powerfully than atomic physics. Crick defined molecular biology as a way of observing "the borderline between the living and the nonliving," that is, the borderline between the scale on which human beings are warm, laughing flesh and the scale on which we are nothing but atoms. A single human thumbtip contains a trillion trillion atoms. The thumb is alive; the atoms are dead. Molecular biologists would explore the acts of life at the level of molecules, atoms joined elegantly together, the smallest working parts in every fingertip and antenna tip. Sydney Brenner defined the new science as "the search for explanations of the behavior of living things in terms of the molecules that compose them." A hybrid science, then, requiring a feel for the behavior of living things, a feel for the behavior of matter, and what Crick calls "the hubris of the physicist."

By the early 1960s the revolutionaries had already done so much with molecules that during a brief interlude of collective depression they decided that their quest was over and they would never find any mysteries equal to the clouds they had just dispelled. Benzer laughs now when he remembers the mood: "We had this feeling that all the molecular biology problems were on the verge of being solved. It was a little bit like the physicists at the end of the nineteenth century saying, 'All we have left to do is one more decimal place.'" As a small boy Crick had worried that the empire of science was expanding too fast. By the time he was grown up there would be nothing for him to do. ("Don't worry, Ducky," his mother had told him. "There will be plenty left for you to find out.") Now the new molecular biologists thought they had done themselves in: there was nothing left. Many of them got as cranky as homecoming heroes the year after coming home. Benzer was still spending his summers at the Marine Biological Laboratory in Woods Hole, but now his work on genes and mutations was so famous that he couldn't walk down Water Street without strangers buttonholing him to tell him about their latest results. At Stony Beach he couldn't get into the water. His cousin Sidney also spent summers at Woods Hole. Sidney's mailbox read "S. Benzer." Someone was always ringing Sidney's bell and asking, "Are you—?"

"No, that's my cousin Seymour!" Sidney would shout and slam the door.

One summer Watson brought a red-hot manuscript to the Benzers' rented cottage at Woods Hole. "He wanted my wife to read it," Benzer remembers with a laugh. "He said, 'These books are bought by housewives, so I want to try it on a housewife.' Of course I read it too. I couldn't put it down." It was Watson's memoir of his discovery with Crick. His working titles included *Honest Jim* and *Base Pairs*. In the manuscript, Watson confessed that he and Crick had peeked at X-ray diffraction pictures of the double helix by their friend Maurice Wilkins and Wilkins's colleague Rosalind Franklin in order to beat them to the discovery of the century. Watson's manuscript was so sensational and for its time so unbuttoned (It began, "I have never seen Francis Crick in a modest mood.") that the Harvard Corporation overruled the editors of the Harvard University Press and ordered them not to publish it.

After the memoir was published, under the title *The Double Helix*, and became a best-seller, Crick brooded and planned his own counter-memoir: "I confess I did get as far as composing a title (*The Loose Screw*) and what I hoped was a catchy opening ("Jim was always clumsy with his hands. One had only to see him peel an orange . . .") but I found I had no stomach to go on."

What was happening in the phage group is what happens in any primate troop. These were the shufflings and scufflings of chimpanzee politics and gibbon gossip after a shift in power. Power was shifting to the new molecular biologists and away from the old biologists. It was clear to many that research with Arrowsmith's purity of intention and Gottlieb's remove from the world was going to be rarer now. *The Double Helix* replaced *Arrowsmith* as the book through which young readers were introduced to the life of science. Watson replaced the Martin Arrowsmith ideal with its opposite: the young scientist who does whatever he has to do to get what he wants, his long hair and his loose shoelaces flying behind him. Of course, the power and pace of the new science itself would have changed the moral climate of the field even without Watson's example. But like the young Arrowsmith, the young Watson would become the standard of a new era—or its harbinger; he spoke to the spirit of the next age.

Crick was sorry to see what Watson regarded as their base behavior immortalized. To Crick, the search had been something more beautiful and interesting than a mere race. "Jim," Crick told one historian. "—The only person who thought it was a race was Jim, nobody else did." He worried that after Watson's book the two of them would be remembered in the friezes and murals of history as young beasts scrambling over other people's backs for a bunch of bananas. The reason *The Double Helix* became a best-seller, Crick wrote, is that it shows that "SCIENTISTS ARE HUMAN, even though the word 'human' more accurately describes the behavior of mammals rather than anything peculiar to our own species, such as mathematics."

EVEN BEFORE the discovery of the double helix, Delbrück had abandoned the search to his followers and gone off alone. Down in his sub-basement laboratory in Church Hall he had begun peering through the microscope at the behavior of single cells, watching *E. coli, Euglena, Paramecium,* and *Rhodospirillum* swim or creep toward light. Delbrück had always preferred to work apart from the crowd. Now he spent hours playing with a fungus called *Phycomyces,* which grows in tiny stalks called sporangiophores—on dung. He was fascinated by the way the sporangiophores of *Phycomyces* are attracted to light, a piece of behavior that is known in biological jargon as phototropism. Over and over he watched the spore-towers grow toward the light—just as Benzer would later watch flies run toward the light in his countercurrent machine. With phage, Delbrück had transformed the study of the gene; now, with *Phycomyces,* he thought he could transform the study of genes and behavior.

When historians look at the great waves of migration from the Old World to the New World, they sometimes speak of the Push and the Pull. Delbrück was feeling both. The old New World was crowded, and he wanted a new New World to live in. He was turning away from the hunt for the gene to ask himself the next great question: How do you go from genes to a living creature that swims, creeps, flies, or grows toward light? What are the atoms of perception? What are the atoms of behavior? These questions were so far ahead of their time that they were guaranteed to get him away from the crowds. Early in 1953, just before the eureka of Watson and Crick, Delbrück dictated a letter to Benzer. Max and his wife, Manny, were sitting in their jeep, "battling our way through the Sunday traffic," he wrote, returning to Pasadena from a four-day desert camping trip in Ensenada, Mexico. He was driving. Manny was taking down the letter on a portable typewriter in her lap: "I am starting on a new venture tomorrow: some experiments on the phototropism of the sporangiophores of *Phycomyces.* If they work, I'll retire from phage." The camping trip was a "vacation before starting a new life."

By the early 1960s, after almost a decade of hard work on *rII,* Benzer was also feeling the Push and the Pull. The gene business was getting so crowded that he thought it would soon be as bad as electronics. Benzer turned forty and looked around him. At Cold Spring Harbor and Woods Hole, more and more toddlers were dragging their diapers in the salt water at low tide. More and more of the fathers and mothers hovering over them were talking about guanine and cytosine, adenine and thymine. Like all parents, they also gossiped about their children's looks and quirks. The discoverers of the gene found themselves repeating the same observations that parents have always traded at the beach: "Just like his father." "Where does she get that?" "Two peas in a pod." "Runs in the family." "Must be in the blood." They pronounced these clichés with arched eyebrows and cosmic winks, they put the old phrases in the goosefeet of quotation marks, to acknowledge that they had already waded out some distance into these mysteries.

Benzer and his wife thought their first daughter, Barbie, was delight-fully lively. They thought their second daughter, Martha, was delightfully calm. Martha had been a different baby from Barbie from her first week in the crib. Watching them play by the water at Cold Spring Harbor or Stony Beach, Benzer wondered, "Are we doing things that differently, or is it genetic?"

"When you have one child, it behaves like a child," Benzer told a lec-ture audience not long ago, accepting the Crafoord Prize in Stockholm for his work on genes and behavior. "But as soon as you have a second child you realize from Day One that this one is different from the other."

Benzer began to notice that whenever he read something about *rII* he felt bored, and whenever he read something about behavior and person-ality he felt alive. He was using what Crick calls the "gossip test." Crick believes that "what you are really interested in is what you gossip about." Benzer listened to his own gossip, his wife's, and his friends', and he felt the pull of the problem of genes and behavior. His friends felt the pull too; the same subject was on all their minds. When Watson became a father, some years later, Benzer had an idea for a present. His wife, Dotty, went out, bought it, and wrapped it. Watson opened the box, and there was a pair of baby sneakers with the shoelaces untied.

Back in 1955, when Benzer had just begun his map of the *rII* gene, Delbrück had predicted that the map would keep him busy for ten years. "He was right," Benzer wrote in 1966, looking back on his *rII* adventure in a volume of reminiscences that he and his friends put together for Delbrück's sixtieth birthday. "In 1965 my interest suddenly turned off post-hypnotically, and it is now more than I can do even to think about the subject." During one single year toward the end of his mapping mania he had published half a dozen papers. That year Delbrück had intercepted a letter that his wife was writing to Benzer's. Max had added a postscript: "Dear Dotty, please tell Seymour to stop writing so many papers. If I gave them the attention his papers *used* to deserve, they would take all my time. If he *must* continue, tell him to do what Ernst Mayr asked his mother to do in her long daily letters, namely, *underline what is important.*"

Suddenly it was hard for Benzer to think of anything worthy of being underlined. He drove out into the cornfields with Dotty again, and again she supported his decision to start over. His work on *rII* had gotten more and more exciting, he later wrote, until it dawned on him how many peo-ple around him were doing the same things, spinning around the same helix or vortex. "I had almost gone down the biochemical drain. Delbrück saved me, when he wrote to my wife to tell me to stop writing so many papers. And I did stop."

TEN YEARS after the discovery of the double helix, Benzer began reading omnivorously: everything he could find about the inheritance of behavior. A few of his phage friends were doing the same thing, including Brenner, Stent, and Delbrück. They were arrogant and confident, but considering what they had accomplished in the previous ten years they had a right to be. Crick considered their hubris "a healthy corrective to the rather plodding, somewhat cautious attitude I often encountered when I began to mix with biologists." Having created a new science, they were racing on. They thought of the problem of instinct as an extension of the problem of inheritance. An instinct, like a gene, is a kind of memory, a gift of time. The gift confers enormous advantages on all those that possess it. We are born knowing a thousand things we could not reinvent in a lifetime if we had to start from scratch. At Caltech, Delbrück used to play chess with the mathematician Solomon Golomb. Delbrück spent sixty minutes to Golomb's one minute and still couldn't win. Delbrück's friends asked him why he kept losing when he gave so much thought to each move. Delbrück said, "I think, but he knows." Now Delbrück and his followers wanted to solve the secret of inborn knowledge the way they had solved the secret of inheritance. They wanted to take instincts apart the way Benzer had taken apart the gene.

What was new in their approach was the attempt to work from the bottom up and from the inside out. Freud worked by introspection, which is looking from the outside inward. For Freud the brain was a black box. When he spoke about the unconscious workings of the mental apparatus, he added a warning for his readers: "I must beg you not to ask what material it is constructed of. That is not a subject of psychological interest. Psychology can be as indifferent to it as, for instance, optics can be to the question of whether the walls of a telescope are made of metal or cardboard." "We must recollect," Freud once admonished his followers, "that all our provisional ideas in psychology will presumably one day be based on an organic substructure." But Freudians and virtually all of the schismatic sects that split away from the Freudians studied the psyche strictly from the top down and the outside in. One psychologist wrote that as far as he was concerned the skull could be full of cotton.

Benzer read these psychologists sardonically. Far back in his mind he could hear Max Gottlieb's asperities in *Arrowsmith*: the true scientist hates "guess-scientists—like these psychoanalysts." Benzer and his circle had found an inner book of symbols that everyone could read, and he hoped their science, unlike Freudian psychology, could build higher and higher on its own foundations.

After Freud, two of the most influential psychologists of the century were John Watson and B. F. Skinner, the founders of behaviorism. Watson and Skinner had decided to proceed not by introspection but by inspection. Science had lifted "the stifling soul cloud" in the study of heaven and earth, John Watson wrote in 1912; now science must lift the cloud from the study of psychology. Watson, Skinner, and their followers tried to see how much they could learn about stimulus and response by conditioning pigeons and rats. For them everything that mattered at the choice points of life came from experience—from environment, from nurture, from outside. John Watson (Watson the behaviorist) made the much-quoted claim that if he were given a dozen healthy babies he would "guarantee to take any one at random and train him" for any job at all, no matter what "his talents, penchants, tendencies, abilities, vocations, and the race of his ancestors."

"Very few people have any notion of the extent to which a science of human behavior is indeed possible," B. F. Skinner wrote in a sort of conspiratorial, movement-starting whisper in 1953, the same year Watson and Crick discovered the double helix. By then the behaviorists' decision to avoid introspection had hardened into the curious view that there is nothing inside us to inspect. Behaviorists had invented a psychology without wants, intentions, or emotions; a psychology—it has been said—without a psyche; a psychology that was all outside. Skinner played with schedules of reward that would get a pigeon to peck at a button. He could space the rewards farther and farther apart and get pigeons to keep pecking until their beaks were worn down to stubs. He was convinced that human beings would have to give up their illusions of mind and soul and emotion, inner life and inner nature, and replace them with stimulus and response. Giving them up once and for all would cure most of the ills of the world, Skinner wrote: "The present unhappy condition of the world may in large measure be traced to our vacillation."

Benzer was not impressed by the behaviorists. Nor did he find what he was looking for in the philosophers; at least not in the kinds of thinkers whom Nietzsche once called with some self-loathing the "knowledge-microscopists," people who cross-examined their own thoughts in their own skulls, asking "What do I know?" and "How do I know it?" and "How do I know that I know?" Benzer's friend Gunther Stent loved to read the philosophers, but most of their circle laughed at them and shook their heads: "They need a little help from their friends."

Every scientist knows the rule called Occam's Razor: Faced with several competing hypotheses, prefer the simplest one. There is also an unspoken corollary that might be called Occam's Castle: Faced with several competing places to build a new science, prefer the simplest one. Pick the place that requires the least preparation, the least digging, hauling out, pouring in, and shoring up. In real estate the rule is location, location, location. In science the rule is foundation, foundation, foundation. Where the foundation is firmest, the castle will rise highest. Where the ground is solid, build there, and the universe is so constructed that you will have a view.

A new science had sprung up on the foundations of physics, chemistry, and Morgan's Fly Room. Benzer and his friends had already added a few stories, always building where the ground was most solid, usually with a sense of very slow and incremental progress, as in the words of the prophet Isaiah: "For precept must be upon precept, precept upon precept; line upon line, line upon line; here a little, and there a little." Molecular biology is Occam's proudest and strangest castle. Rising from what is (from the point of view of real estate) one of the world's unlikeliest places, the Fly Room, it has grown into the single most towering accomplishment of the human mind in the last one hundred years.

Benzer wanted to build on that foundation. He was convinced that there must be genetic differences behind the innumerable quirks of our bodies and minds, and he was sure that these differences must matter at every turn of our behavior, at every one of our choice points. He wanted to find some of these genes and figure out how they make a difference. In those years these were new questions, and their strangeness attracted Benzer like filet of snake. He went into his new problem as if he were stepping into the dark, not sure how many steps there were or if there were any steps at all.

On and off during the first half of the 1960s Benzer read and mused, visiting laboratories, taking courses, looking for a place to start. His bookshelf from that period of his life includes titles ranging from Darwin's *Expression of the Emotions in Man and Animals* and Galton's *Hered-*

*itary Genius* to *The Machinery of the Brain, Physiological Psychology, The Physiological Clock, Behaviorism, Behavior of the Lower Organisms,* and *ABC and XYZ of Bee Culture* (33rd edition).

Everyone since the beginning of human thought had been looking for a solid foundation. Socrates in one of his last dialogues before drinking the hemlock complained that arguments about human nature "visibly shift their ground instead of keeping still." He cried, "Don't you see that our discussion has gone right round and come back to the point from which we started?"

Pascal wrote, "We burn with desire to find a firm footing, an ultimate, lasting base on which to build a tower rising up to infinity, but our whole foundation cracks and the earth opens up into the depth of the abyss."

Darwin wrote in his secret notebook, "To study Metaphysics as they have always been studied appears to me to be like puzzling at astronomy without mechanics." In other words, to approach the great metaphysical questions without understanding how the mind works, without getting inside the anatomy and mechanics of the mind, is an enterprise as hopeless as to approach the motions of the stars and planets without understanding celestial mechanics. "—Experience shows the problem of the mind cannot be solved by attacking the citadel itself," Darwin scribbled in his notebook. "—The mind is function of body.—we must bring some *stable* foundation to argue from."

Genes would be Benzer's foundation stones. He wanted to build from this foundation to places no one thought the human mind could go. He wanted to find something new about some of the oldest cornerstones of human experience—time, love, and memory—and the oldest questions of heredity, nature, and nurture. He thought about the question at all hours in his physics laboratory at Purdue, and he thought about it in the summers at Cold Spring Harbor, where Dotty always made their rented house a Brooklyn away from Brooklyn. Max and Manny Delbrück, strolling there for dinner, would smile to see Dotty waving at them from the porch amid blowing laundry. Dotty was as down to earth as Seymour, and she anchored him the way Leora anchored Martin Arrowsmith.

Natural philosophers had struggled forever with nature and nurture, nurture and nature, wheeling and wheeling in the fugue of the questions. Their writings were a fumbling of wings in a locked room: whole generations trapped like flies in a fly bottle, or like the bat in D. H. Lawrence's poem, going

*Round and round and round*
*with a twitchy, nervous, intolerable flight*

Benzer wanted to work from the gene to the neuron to the brain to behavior, and he hoped to dissect them all the way he had dissected the gene. While he thought and read, he asked Dotty to buy brains at the butcher's shop: sheep, cow, goat, pig, and chicken brains. One by one she brought them home, and one by one he dissected them, usually in the middle of the night. Afterward, he ate them.

# PART TWO

# Konopka's Law

*Things are always best in their beginning.*

—Blaise Pascal,
*Lettres Provinciales*

CHAPTER SIX

# First Light

*I'll tell you how the Sun rose*
*A Ribbon at a time.*

—EMILY DICKINSON

*Everyone who ever lived . . . lived at a moment of equal*
*astonishment.*

—RICHARD POWERS,
*Galatea 2.2*

BENZER FOUND his next destination in a little book called *The Machinery of the Brain,* by Dean E. Wooldridge, in which a drawing of the furrowed lobes of the human brain loomed attractively on the title page like attainable mountains.

In *The Machinery of the Brain,* Benzer read about the early experiments of a biologist at Caltech named Roger Sperry. Sperry had cut the optic nerves of a toad and reconnected the left optic nerve to the right half of the toad's brain and the right nerve to the left half of its brain. Toad nerves can be cut and spliced like wires; once human nerves are severed, unfortunately, they will not reconnect (although the science that Benzer helped to start may change that yet). An optic nerve is actually a bundle of tens of thousands of nerves. These fibers cross and twist and wind around one another as if none of them had been sure when they first grew in the embryo that they knew the way between the eye and the brain. Obviously they do find their way in the embryo. They cannot find their way again in an adult human being, since the nerves will

73

not grow; but Sperry wanted to see if they could find their way a second time in a fully grown toad.

A few weeks after the operation, Sperry was amazed to find that the toad acted like a normal toad. When a fly came down within its reach, the toad darted out its tongue. Somehow those tens of thousands of fibers in each optic nerve had found their way, and the toad could see again. The toad had only one problem: if a fly came from the right, it stuck out its tongue to the left, and if a fly came from the left, it stuck out its tongue to the right.

In *The Machinery of the Brain,* Benzer also read about Sperry's experiments with cats. The optic nerves from a cat's right and left eyes meet briefly on their way to the two hemispheres of the brain and then part again. At their meeting point, the optic chiasma, they share some of their information. Human beings have an optic chiasma too. With a scalpel, Sperry cut through the optic chiasma. Again he let his subject recover from the operation. Now he led the cat to a choice point, the way Benzer would later do with his own subjects in the countercurrent machine. There might be two doors, one marked with a circle and the other with a square. The cat would study these doors through one eye—Sperry had given it an eye patch—so the image of the circle and the square would travel from one of the cat's eyes to one half of the cat's brain. And after a few tries the cat would learn to choose the door with the circle because that one led to a pellet of food.

The cat could learn this kind of lesson through either eye, and it would choose the circle again the next time even if Sperry switched the eye patch. Even though one eye had sent its messages to one hemisphere, both hemispheres had somehow learned the lesson. Sperry assumed that the lesson had been shared across the corpus callosum, the thick bundle of nerve fibers that connects the hemispheres of the brain in cats and in human beings. He cut the corpus callosum; and when he switched the cat's eye patch, the cat did not know which door to choose, the circle or the square. If the left brain had learned the lesson, the right brain had not; and if the right brain had learned the lesson, the left brain had not. Sperry could even teach the left brain to choose the circle and the right brain to choose the square, and the door the cat chose would depend on which eye saw the choice point.

Sperry repeated this experiment with a monkey, cutting the optic chiasm and the corpus callosum. This time he also gave one half of the monkey's brain a frontal lobotomy. He put an eye patch on the monkey and showed it a snake. Monkeys have an instinctive fear of snakes. If the

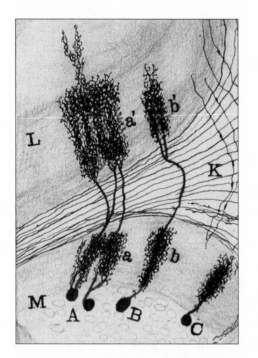

*In the early years of the twentieth century, the pioneering Spanish neu-
roanatomist Santiago Ramón y Cajal was the first to stain, draw, and
number individual nerve cells in the brain. (Here, nerves from the brains
of a bee and a fly.) In the early 1960s, Seymour Benzer and other molecular biologists studied Cajal's wiring diagrams and wondered if they could
now trace the connections between genes, nerves, and behavior.*

sight of the snake went to the undamaged half of its brain, the monkey
screamed, defecated, and tried to escape. But if the sight went to the
lobotomized hemisphere, the monkey gave the snake a "What, me
worry?" look. It was as if the monkey had two separate brains, or as if
there were two monkeys in one body.

*The Machinery of the Brain* gave Benzer the same sense that *Arrowsmith* had given him as a teenager in Brooklyn and *What Is Life?* as a
young physicist in Lafayette: the sense of having discovered a personal
road map. The author, Wooldridge, had once been in charge of research
and development for the Hughes Aircraft Company. He was a founder
and president of a major high-technology corporation, Thompson Ramo
Wooldridge. Wooldridge had quit to hang around laboratories like
Sperry's at Caltech. He wrote that he meant his book to be "a travelogue—a description of an exotic land by one who visits it for the first

time." He hoped it would serve another lapsed physicist somewhere as a point of departure "for more intensive studies."

In 1965, Benzer took a year's leave of absence from Purdue and visited Sperry's laboratory on the third floor of Church Hall. Delbrück was there, too, down in the basement, working on the fungus that grows toward light.

In Sperry's lab, Benzer watched biologists study the brains of goldfish, monkeys, frogs, chicks, chameleons, cats, and human beings. The work on human beings later won Sperry a Nobel Prize. He had begun to study epileptics whose brains' right and left hemispheres had been surgically severed. Cutting the corpus callosum prevents epileptic seizures from spreading from one side of the brain to the other. Surgeons had been performing these operations for years in the belief that their patients' behavior would not be affected. One authority on the brain joked that the only role of the human corpus callosum seemed to be to pass epileptic seizures from one side of the brain to the other. Another authority said that its only role was to keep the two halves of the brain from sagging.

But Sperry showed that under special conditions the behavior of split-brain patients is bizarre. A hybrid face is flashed on a screen: One side is a man's, the other side a woman's. The patient is presented this image in such a way that his left brain sees only the man's face and his right brain sees only the woman's. If the patient is asked to report what he just saw he will say, "A man," because the left hemisphere is dominant in language. But if the patient is asked to point to a face that matches the one he just saw, he will point to the woman, because the right hemisphere is dominant for action and movement. Now the word "WALK" is flashed on the screen in such a way that only the patient's right brain sees it. Soon afterward he gets up and begins to walk away. The experimenter asks him why he got up. "I'm going to get a Coke," he says. He cannot explain the real reason because only his right brain knows, and his right brain is mute.

The human brain, like the brain of a monkey, a cat, even a toad or a fly, is made of many separate working parts. Today some neuroanatomists speak of modules. Just as each lobe of the human brain has its own characteristic infoldings, each lobe and each region of the brain and brain stem has its own characteristic function: there is no breathing without the medulla oblongata, no sensation of smooth movement without the pons, no eye movement without the midbrain. The modules may be even finer-scale than that, as Sperry and others were beginning to discover. This is the essence of the idea that Freud and his followers were exploring from the outside by introspection: that we have conflicting motives and drives, and that we are aware of only a small subset of them at a time. Sensitive human beings have always known this and have felt the strain of their own incongruities. Henry David Thoreau writes in his poem "Sic Vita" (Such Is Life):

> *I am a parcel of vain strivings tied*
> *By a chance bond together,*
> *Dangling this way and that, their links*
> *Were made so loose and wide.*

We have all had occasion to feel that the right hand does not know what the left hand is doing. We made a resolution not to eat or drink, and then we watched the hand pour or eat—while we invented stories to account for and explain the act to ourselves, like Sperry's subject on his way to the Coke. Each of us is a loose parcel of strivings that normally display themselves to us not all at once but across time. Some of these drives seem to visit us from a long way back in time, and some threaten to wreak havoc with the rest. A life is a great parliament of instincts, as Konrad Lorenz once put it. This parliament is what Benzer proposed to explore through the genes.

IN SPERRY'S LABORATORY, Benzer prowled from bench to bench. He knew that phage had spoiled him. With phage he could hold billions of test subjects in a dish no bigger than the palm of his hand. He could breed dozens of generations in a day and find one freak in a billion at a glance. After enjoying that kind of speed, convenience, and instant gratification, he could not imagine trying to breed and cross goldfish, monkeys, frogs, chicks, cats, or chameleons. And as he said later, drily, "Humans were ruled out because it is so difficult to convince them to mate in the right combinations, generations take too long a time, and the offspring are too few."

Benzer considered ants, spiders, and bees. Bees interested him, but unlike Mendel he did not love the idea of building hives and mating chambers. For a while he kept spiders spinning webs in jars, admiring their wonderful fixed-action patterns. But he thought spiders would be almost as inconvenient for his purposes as bees. "I searched the literature," he says. "No one had ever done a Mendelian cross between two spiders. One reason was that the females ate the males."

While he thought about his problem, he kept reading voraciously. He admired *The Expression of the Emotions in Man and Animals,* which Darwin had published in 1872. Those were the days when the sun never set on the British Empire, and Darwin had sent out a questionnaire to explorers, missionaries, and "protectors of the aborigines" in every corner of it:

(1.) Is astonishment expressed by the eyes and mouth being open wide, and by the eyebrows being raised?

(2.) Does shame excite a blush when the color of the skin allows it to be visible? and especially how low down the body does the blush extend?

(3.) When a man is indignant or defiant does he frown, hold his body and head erect, square his shoulders and clench his fists?

(4.) When considering deeply on any subject, or trying to understand any puzzle, does he frown, or wrinkle the skin beneath the lower eyelids?

And so on. The replies that Darwin received from Australia, New Zealand, India, Africa, the Malays, China, and the American Northwest were all affirmative. People's expressions are broadly the same all over the planet, which is one reason that Hollywood's empire is at least as large today as the British Empire was in 1872.

The fundamentals of human expressions are inherited, and Benzer was sure there was much more to explore. The flinch reflex, for instance, is obviously adaptive and hard-wired. Many of us flinch when we see a snake, and some of us react like Sperry's monkey. Again the reaction is probably hard-wired and adaptive—at least, it was adaptive for millions of years of human evolution until we got out of the trees and into the cities. We have been living in cities for only a short fraction of the life of our species, and evolution has only partly dampened the reflex in most of us, like a partial lobotomy. In *The Expression of the Emotions,* Darwin describes an experiment at the London Zoo: "I put my face close to the thick glass-plate in front of a puff-adder in the Zoological Gardens, with the firm determination of not starting back if the snake struck at me; but, as soon as the blow was struck, my resolution went for nothing, and I jumped a yard or two backwards with astonishing rapidity. My will and reason were powerless against the imagination of a danger which had never been experienced."

Benzer also enjoyed reading books by Darwin's cousin. Galton did bizarre introspective experiments to get at the bits of inheritance that he was convinced must underlie behavior. By exploring his own mind and quizzing others, he explored the ways we are all alike and the ways we differ. Not everyone answered Galton's questionnaires. Darwin refused ("I have never tried looking into my own mind."). Still Galton found that most people seemed to enjoy introspecting. "I think that a delight in self-dissection must be a strong ingredient in the pleasure that many are said to take in confessing themselves to priests."

For instance, Galton corresponded with a calculating prodigy who was the son of a calculating prodigy. The man told him that he always saw numbers in his mind's eye, and he sent Galton a drawing. "It began," Galton writes, "with the face of a clock, numbered I to XII, and then tailed off, much like the tail of a kite, into an undulating curve, having 20, 30, 40, etc., at each bend." Galton was surprised to discover that these number visions are fairly common. When he asked about them in a large lecture "up went a multitude of scattered hands all about the body of the hall." And number forms seemed to run in families. They reflected something about the quirky individual construction of individual minds, and the quirks could be traced—as we say now—to genes. Years later Benzer would do an informal survey and discover number visionaries in his wife's family.

Time and again Galton's experiments gave Benzer glimpses of what Galton called "the number of operations of the mind, and of the obscure depths in which they took place, of which I had been little conscious before." It was like walking down into the basement when the plumber was there, Galton wrote prosaically, and seeing for the first time "the complex system of drains and gas and water pipes, flues, bell-wires, and so forth, upon which our comfort depends, but which are usually hidden out of sight, and with whose existence, so long as they acted well, we had never troubled ourselves." And what a difference between the view of the people of the house and the view of the plumber!

"Laypeople—I don't mean very sophisticated people, but ordinary laypeople—think if something is natural it doesn't require an explanation," says Francis Crick. "You know: 'What is there to worry about? It's perfectly natural!' You see? Whereas we know that some things that are natural are often extremely intricate in the mechanisms that are needed to *make* this natural behavior—as we soon find out if something goes wrong. But that's the reaction of many laypeople, that they think their behavior is basically simple. They do it, it's perfectly natural, what is it to explain? And to think that it's due to genes, or something else, that's

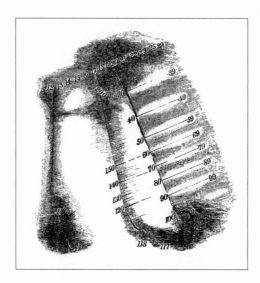

*Darwin's cousin Francis Galton was convinced that each mind has quirks and that many of the quirks are inherited. Galton reported that one out of every fifteen women and one out of every thirty men sees what he called "number-forms" in the mind's eye. Whenever these number visionaries think of a number, they always see it floating in the same place in imaginary space. One subject told Galton that he always saw the number one floating low and to the left, the number one hundred floating low and to the right, with every digit in between forming a complicated spectral arc. Number forms like these may run in families.*

really *terrible.*" Crick laughs. "Although all parents will remark to you how different their children are. And they notice it from an extremely early age."

Galton knew nothing about the genes that underlie all these operations of the brain. But Benzer was sure that Galton was right: There are bound to be genetic differences for a thousand mental operations and a thousand individual quirks too. Galton urged future generations of scientists to study the variability of human instincts. Like Darwin, he singled out the fear of snakes. "I myself have a horror of them," Galton wrote, "and can only by great self-control, and under a sense of real agitation, force myself to touch one." Sometimes Galton forced himself to watch rabbits and birds being fed to the snakes at the London Zoo. He found the sight horrible but fascinating, and he could not understand

*"When you have one child, it behaves like a child," Benzer says. "But as soon as you have a second child you realize from Day One that this one is different from the other." The Benzer family's "Season's Greetings" photograph, 1954.*

how children and their nurses could stand right next to him and smile or stroll on by. "Their indifference was perhaps the most painful element of the whole transaction. Their sympathies were absolutely unawakened."

Galton also had a horror of blood, and he thought that this might be another human instinct: "but I have seen a well-dressed child of about four years old poking its finger with a pleased innocent look into the bleeding carcase of a sheep hung up in a butcher's shop, while its nurse was inside." Galton thought schoolteachers should take surveys of children's nightmares and goose bumps: "It would be necessary to approach the subject wholly without prejudice, as a pure matter of observation, just as if the children were the fauna and flora of hitherto undescribed species in an entirely new land."

Benzer knew he had to start with something simpler.

WHENEVER HE WANDERED out of Sperry's laboratory, he would pass the relics of Morgan's last Fly Room. Morgan's collected works, beginning with his first notes on mice and starfish, reposed in rows of file cabinets up and down the corridor. Morgan himself was long since dead, but Alfred Sturtevant, the old veteran of Morgan's Raiders, still came around often, a pipe clenched between his teeth, to see what was new and interesting in genetics, and to tend an experimental bed of irises that he had planted just outside the building. One of Sturtevant's best students, Ed Lewis, had inherited his old lab space on the third floor, along with thousands and thousands of mutant flies—and of course the milk bottles.

Benzer watched Lewis sort flies, and Lewis watched Benzer watch him. Since the discovery of the double helix, the gulf between molecular biology and the rest of biology had widened every year. Molecular biologists were known as bad boys, bullyboys, arms of the flood. At Harvard, Jim Watson was trying with enormous energy and without enormous tact to pack the biology department with molecular people and get rid of all the deadwood: the field biologists, taxonomists, ecologists, ethologists, and naturalists. Among the young professors who arrived at Harvard at the same time as Watson was E. O. Wilson, who would become, among other things, one of the century's great field biologists, taxonomists, ecologists, ethologists, and naturalists. At one faculty meeting Wilson proposed hiring another ecologist. He heard Watson say softly, as if to himself, "Are they out of their minds?"

"What do you mean?" Wilson asked.

"Anyone who would hire an ecologist is out of his mind."

Wilson had read *What Is Life?* in college at the University of Alabama in Tuscaloosa, and he had been just as thrilled as Watson, Crick, and Benzer ("Imagine: biology transformed by the same mental effort that split the atom!"). Both Wilson and Watson would end their careers convinced that the search for the atoms of behavior is the central quest of science. But in those days, Watson had nothing to say to Wilson when they passed in the halls at Harvard, even when they were the only two people in the hall. Wilson's memoir *Naturalist* includes a chapter "The Molecular Wars." The chapter begins, "Without a trace of irony I can say I have been blessed with brilliant enemies." James Dewey Watson was one of them. "When he was a young man, in the 1950s and 1960s," Wilson writes, "I found him the most unpleasant human being I had ever met."

Not since the Age of Enlightenment had the world seen such a crew of intellectual cutthroats, divinely assured of their rights of succession and their place in history. The philosophes of the Enlightenment also had their share of tall, thin, prognathous young men, and many of their contemporaries found them (in the words of Horace Walpole) "solemn, arrogant, dictatorial coxcombs—I need not say superlatively disagreeable."

Seymour Benzer was not an unpleasant human being, but he was one of the revolutionaries. In Church Hall he and Delbrück and sometimes Watson (when he came to visit) strode through the corridors talking about events "in the days of genetics" as if those days were ancient history—even though Ed Lewis was still sitting in his lab crossing mutant flies, and the man who had made the first map of the genes was kneeling just outside the building, weeding his irises. Someone in the public relations department at Caltech once interviewed Benzer as part of an in-house oral history project. She asked Benzer if Lewis had been held in contempt by the new breed of molecular biologists in the 1960s. "No. He was a nice guy," Benzer replied. "He was very good with flies. But at the time it seemed sort of like having a Greek mythology scholar: it's nice to have one around for the university at large. He taught the genetics class, and kids counted the flies. Of course I'm giving you the jaundiced point of view. He was the true inheritor of the Morgan-Sturtevant tradition, and that was just fine."

Lewis himself felt his isolation painfully. "*Drosophila* went into almost total eclipse," he says. "Delbrück would pound the table: 'Genetics is dead! Genetics is dead! Genetics is dead!'" Over and over, Delbrück said it in so many words: Molecular biology is the only biology.

(Many years later, around the corner from the campus, sitting in Max Delbrück's favorite old rocking chair, Manny Delbrück would giggle when she remembered the apocalyptic speeches her husband used to make. "You see," she said, "Max didn't *know* any other biology.")

As it turned out, what Lewis was learning in his Fly Room would someday thrill a new generation of molecular biologists and would win him a share of a Nobel Prize. But at that time, Lewis saw no point in trying to convince Delbrück that what he was doing was interesting. Lewis was smaller, gentler, and quieter than most of the bullyboys. He liked to keep tanks of rare tropical fish and anemones in his laboratory, and he raised generations of octopi, which Benzer, who had never seen an octopus embryo, found supremely beautiful. Lewis had long, owlish eyebrows like nerve endings looking for a home. And Lewis was an owl, like Benzer himself. Sometimes just before dawn, when Benzer roamed the corridors thinking about genes, nerves, and behavior, he would hear, through Lewis's closed door, the sounds of a flute.

BENZER KNEW what he wanted: He wanted to go from the gene to the kinds of instincts that ethologists studied out in the wild. Ethologists studied pieces of animal behavior as pieces of anatomy, portions of our inheritance that have evolved over thousands of generations just like a thorax or a pelvic bone or a braincase. When they studied the imprinting of goslings on geese, the mating dances of ducks and sticklebacks, and the massive collective engineering projects of ants and bees, they tried to break apart each act of instinctive behavior into a series of steps, the steps they called "atoms of behavior." Of course, most ethologists were field biologists. They worked outside, watching bees, dragonflies, greylag geese, and ruddy Egyptian ducks, as Konrad Lorenz describes in his memoir *King Solomon's Ring*. They loved to tramp along streams and riverbanks watching their subjects in action, and a few of them learned to talk in so many languages of duck and goose that sometimes in the heat of the moment one of Lorenz's field assistants would get mixed up: "*Rangangangang, rangangang*—oh, sorry, I mean—*quahg, gegegegeg, Quahg, gegegeg!*" They ignored both genetics and molecular biology. They assumed that variations in behavior do pass down from generation to generation, and they left it at that. "Atoms of behavior" was only a metaphor. They studied instincts from the outside.

As one of the world's first molecular biologists, Benzer thought he could do what no one else had done: look at the actual atoms. But after his ten years of hard labor mapping the gene *rII*, he did not want to have to start mapping genes again. He wanted to get straight to the fun and study the behavior of an animal whose genes had already been mapped.

More and more often, Benzer found himself strolling out of Sperry's lab in the middle of the night, sometimes with a book in his hand, and dropping by Lewis's Fly Room to look at the coral fish and the baby octopi. Whenever he dropped in, Lewis would be busy inspecting mutant flies through the microscope and adding their mutations to the chromosome maps that his teacher Sturtevant had begun in the middle of the night back in 1911.

If Benzer was ever going to disentangle nature and nurture, it seemed self-evident to him that the thing to do was to keep the environment constant and change the genes. He knew that the maps on Lewis's walls were still the most extensive genetic maps of any organism in the world. If he worked with them, he would be making a kind of splice between the old and the new, just as he had done in his adventures with *rII*; and he could ask for help from Sturtevant and Lewis, two of the last Lords of the Flies.

Of course, his old mentor and Zen master Max Delbrück would cackle. But Max himself had a saying: "Don't do fashionable research."

One night Benzer stopped by Lewis's laboratory and borrowed a milk bottle full of fruit flies. He also borrowed some test tubes—he couldn't find a test tube anywhere in Sperry's lab. He put a lightbulb on his benchtop, held two test tubes mouth to mouth, turned out the overhead lights, and watched a fly run toward the light.

# CHAPTER SEVEN

# First Choice

*The brain is so vigorous and active it insinuates itself into all places and times; reaches the heights, searches the depths, peers into all those recluded cabinets of nature wherein she hath stored up the choicer and abstrussest pieces of all her workmanship, and these it contemplates and admires.*

—NATHANIEL WANLEY,
*The Wonders of the Little World,* 1788

*I thought of a maze of mazes, of a sinuous, ever growing maze which would take in both past and future and would somehow involve the stars.*

—JORGE LUIS BORGES,
*"The Garden of Forking Paths"*

IN A LABORATORY seminar in 1966, Benzer told Roger Sperry's students how he had trapped a fly in a test-tube tunnel, with a light at the end of the tunnel. He told them that most of the flies in the tunnel had moved toward the light most of the time—instinctively, the way a moth flies toward a candle flame. He told them that with a series of test-tube tunnels he might be able to separate the light-lovers from the dark-lovers, just as a chemist working with molecules can separate the water-lovers from the oil-lovers.

This countercurrent machine would be just the beginning, Benzer said. It would be the prototype for a whole series of experiments in which he would look for mutant instincts and mutant behavior just the way Sturtevant and Ed Lewis looked for mutant wings and abdomens. Like Sturtevant and Lewis, Benzer would find plenty of mutants, because he would feed his flies a mutagen. Lewis recommended a poison called ethyl methane sulfonate (EMS), a mutagen that he had popularized in Fly Rooms around the world. X rays tend to remove huge chunks of DNA with a single hit—anywhere from a thousand to a million letters. But EMS is kinder and gentler (like Lewis himself), and it will generally change only one letter of the genetic code at a time. By using EMS, Benzer would multiply the chances of finding interesting mutants and interesting behavior.

Finally, Benzer explained, he would be able to work much faster than Morgan's Raiders with their jeweler's loupes and microscopes. With a countercurrent machine, instead of examining flies one at a time, he could test a hundred flies at once. The experiments would be very simple and quick, like his phage work. In two minutes he could get as much statistical information as a behaviorist could get in several months of work with rats. He could crack the problem of genes and behavior wide open.

"I laid out this whole plan," Benzer remembers now. "The whole lab was in a sort of uproar for about a week after that, people arguing with each other. They were pretty much split down the middle between those who thought this was great stuff and others who thought this was pure crap, that I'd never solve any problems that are important. They were really screaming at each other." It was as if Sperry's students were stuck in a countercurrent experiment. "Science as something already in existence, already completed, is the most objective, impersonal thing that we humans know," Albert Einstein once said somewhere. "Science as something coming into being, as a goal, is just as subjectively, psychologically conditioned as are all other human endeavors." Benzer remembers the reaction in Cold Spring Harbor in June 1953, when Watson first presented the structure of the genetic material. The announcement had not even been on the program; it was inserted as a special lecture. After the lecture some people were literally jumping up and down. Other people were saying, "So, big deal." "Double helix—so what?"

(The next speaker had a hard act to follow. It was Max Delbrück, announcing his entry into the field of genes and behavior with a talk entitled "Phototropism in Fungi.")

In the Sperry lab, the scientists who hated Benzer's idea really hated it. They studied nerves and brains. Why look at genes if you are interested in nerves and brains? "Obviously, that attitude was completely

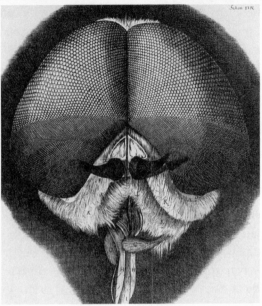

*When the world's first molecular biologists turned to the study of genes and behavior, their two most successful laboratory animals—worms and flies—were old favorites of the microscopists. These drawings of nematode worms and the head of a fly are from Robert Hooke's* Micrographia, *published in 1665.*

wrong," Benzer says now, "because genes are what make all the parts of a nerve. It was obvious to me."

"Well, of course they came out of a totally different tradition in the Sperry lab," explains the *Drosophila* geneticist Michael Ashburner, who was working in Church Hall at the time. "The idea of using genetics as a tool, rather than a dissecting needle!" He laughs. Benzer's plan was bound to raise hackles in a laboratory like Sperry's, he says. "*A*, of course, it was only insects. Yeah? And *B*, they would probably have had little idea or appreciation of how powerful genetic analysis could be."

Either reason would have been enough to turn people off. Many biologists regarded molecular biologists as coldly as the molecular biologists regarded them. In most universities they were locked in the kind of ritual tribal hostilities that E. O. Wilson speculates may be instinctive. That is how Wilson describes the molecular wars at Harvard: "At faculty meetings we sat together in edgy formality, like Bedouin chieftains gathered around a disputed water well."

But even molecular biologists found Benzer's idea new and strange at the time.

Then there was the bug factor. So many people are so revolted by bugs that this reaction may be instinctive too, along with our fear of snakes. Darwin once wondered if monkeys' terror of snakes explains their "strange, though mistaken, instinctive dread of innocent lizards and frogs. An orang, also, has been known to be much alarmed at the first sight of a turtle." Likewise, many human beings have an instinctive loathing of spiders, and maybe some of us carry that over to the innocent fruit flies. "What sort of insects do you rejoice in, where you come from?" the Gnat asks Alice in *Through the Looking Glass*. "I don't rejoice in insects at all," Alice explains. Of course, some scientists do rejoice in them. Darwin's first passion as a field biologist was beetles. Mendel bred bees after peas, and he had his bees and beehives painted on the chapel ceiling. The ethologist Karl von Frisch called his bees "my magic well." E. O. Wilson, when asked what to do about ants in the kitchen, replies, "Watch where you step." Many biologists who are drawn to pioneering work, drawn to the fringe, are also drawn to organisms that everyone else avoids. As Monod used to say at the Institut Pasteur when his students worried about working on lowly viruses and bacteria, "Remember, there is always plenty of room at the bottom."

Listening to Roger Sperry's students snort and snicker about flies, Benzer remembered a story he had heard back in his physics days. The head of Purdue's secret radar laboratory, Karl Lark-Horovitz, had been one of the first physicists to realize that one might use radioactivity to trace the inner workings of living things. Before the war, Lark-Horovitz had given a lecture on the subject in Vienna, and a woman had come up to him afterward. "Dr. Horovitz, this is fantastic," she said. "To even give an enema to a cockroach is already a great achievement. But to use radioactive phosphorous is the height of sophistication."

When Benzer's mother arrived from Brooklyn for a visit and heard what her only son, her only college graduate, was planning to do now, she said, "From this you can make a living?" She took his wife aside. "Tell me,

*"Go it Charlie!" Darwin rode into biology on the back of a beetle. When he was a student in Cambridge, beetle hunting was one of the only pursuits he took seriously. These sketches were drawn by a fellow beetle hunter, Albert Way.*

Dotty, if Seymour's going to examine the brain of a fly, don't you think we should have *his* brain examined?"

"Well," Francis Crick says today, laughing, "most laypeople are astonished that one studies *Drosophila* at all. I mean, that's the thing you find if you talk to laypeople: 'Why should it be of interest?' You see. And that was always said in the days of genetics. Whereas we know they have been very *interesting* in genetics."

THERE WAS ONE more reason why Benzer's idea sounded outré in 1966. He was not only going back to flies; he was also going back to Galton, or at least to a line of research that Galton had started; and in the United States in the 1960s, Galton was anathema.

Galton had made the inheritance of behavior the first great sustained study in the science now known as genetics because he wanted to breed better human beings. That is why he had been so interested in the idea that inheritance comes in bits: he wanted to rebuild the human race particle by particle. He thought the construction of our bodies and minds must be like the construction of houses he had seen in Italy, many of which are built from pieces of older houses that had been pillaged or torn down. In the facades of these houses, Galton had often

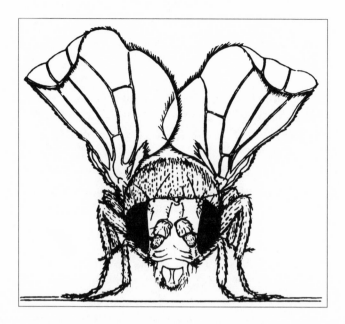

*Full-face portrait: A bent-wing mutant from the first Fly Room at Columbia at the turn of the century. "You can see it is certainly a very thoughtful and kind animal when you get to know it close up," Benzer says in lectures. "What is behind this facade is actually a very compli-cated brain."*

noticed a column or a lintel that had been recycled, sometimes bearing fragments of inscriptions from the house before or the house before that. Likewise, in our human inheritance, he wrote, everything comes from the past, lintel from lintel, column from column, chunk of wall from chunk of wall.

Pushing this metaphor one step further ("which is as much as it will bear," Galton wrote), he imagined that it might explain the strange play of family resemblance in looks and behavior. "Suppose we were building a house with second-hand materials carted from a dealer's yard," he wrote,

> we should often find considerable portions of the same old houses to be still grouped together. Materials derived from various structures might have been moved and much shuffled together in the yard, yet pieces from the same source would frequently remain in juxtaposition and may be entangled. They would lie side by side ready to be carted away at the same time and to be re-erected together anew. So in the

process of transmission by inheritance, elements derived from the same ancestor are apt to appear in large groups, just as if they had clung together in the pre-embryonic stage, as perhaps they did.

This is one of Galton's many visionary passages. He is sketching the principle that later allowed Sturtevant to make the first map of the genes on a chromosome: *A, B, C, D, E.* It is the same principle that allowed Benzer to make the first detailed map of a gene's interior, dissecting a single one of Galton's bits; and Benzer would come back to this principle again.

For Galton, all of this science was part of a utopian dream. At first he called it "viriculture," meaning the cultivation of men. Then he hit on the Greek word *eugenēs,* "namely, good in stock, hereditarily endowed with noble qualities." And the first premise of eugenics was the link between genes and behavior. "We must free our minds of a great deal of prejudice before we can rightly judge of the direction in which different races need to be improved," he wrote in the opening pages of his *Inquiries into Human Faculty* in 1883; but having said that, he felt "justified in roundly asserting that the natural characteristics of every human race admit of large improvement in many directions easy to specify." Everyone knew, for instance, that women are "capricious and coy"—airheads. Everyone knew that Jews are double-dealing misers. And so on.

Galton, who outlived Darwin by decades, was thrilled to read the papers in which Morgan and his Raiders made genes real at last. The discoveries that poured out of the Fly Room in the first decades of the twentieth century did wonders for the Eugenics Society that Galton had founded in London. Morgan's flies helped Galton win streams of influential gray-haired converts to the cause. Morgan himself wanted no part of eugenics ("A little goodwill might seem more fitting."). But Muller, the most visionary of Morgan's Raiders, became an ardent eugenicist from his first days in the Fly Room. Later on, when Muller discovered how to make mutant flies with X rays, he predicted that his transformation of flies would lead to the transformation of the human species.

Given the prejudices of Galton's time and class—the assumption that there really are superior and inferior breeds of people, races as different as foxes and hounds or weeds and roses—one can understand how eugenics might have seemed to Galton a beautiful dream. He was very proud when a botanist named a genus of flowers after him ("a whole genus of flowers of singular beauty"), and he put a little picture of *Galtonia candicans* at the bottom of the last page of his memoir, *Memories of My Life,* along with a few inspirational lines about eugenics. He hoped the human race would be improved and beautified too. Someone once lamented to Galton (perhaps with an irony he missed) that there would be no room left in his perfected world for pity. Precisely so, Galton said: "But it does not seem reasonable to preserve sickly breeds for the sole purpose of tending them, as the breed of foxes is preserved solely for sport."

In the United States, Galton's books started a vogue for eugenics and even a vogue for the name Eugene. Sinclair Lewis satirized the American movement in *Arrowsmith* with his description of a typical midwestern "Health Fair" in which the chief booth was occupied by the Eugenic Family: "They were father, mother, and five children, all so beautiful and powerful that they had recently been presenting refined acrobatic exhibitions on the Chautauqua Circuit. None of them smoked, drank, spit upon pavements, used foul language, or ate meat." While the young Martin Arrowsmith works another booth, answering the public's questions about bacteria and hygiene, a detective recognizes the Eugenic Family as the Holton gang ("The man and woman ain't married, and only one of the kids is theirs. They've done time for selling licker to the Indians.").

Eugenic Fairs and mass sterilization programs in the United States helped inspire the Nazis in Germany. Two months after they came to power in 1934 the Nazis passed the Law for the Prevention of Genetically Diseased Progeny, which mandated sterilization of the congenitally feebleminded, along with anyone suffering from deformities, epilepsy, manic depression, Huntington's chorea, hereditary blindness or deafness, even alcoholism.

In London in 1936, Julian Huxley, a brother of Aldous and a grandson of Thomas Henry Huxley, Darwin's bulldog, gave the keynote speech at the Galton Dinner in the Waldorf Hotel. Huxley called eugenics "one of the supreme religious duties," almost a platitude by then, and he illustrated it with a mutant fly, *abnormal abdomen*. After Huxley's speech, the president of the International Union for the Scientific Investigation of Population Problems, Colonel Sir Charles Close, got up and applauded Huxley for a talk so full of valuable remarks: "We cannot digest all of them at the present time; to attempt to do so might bring us to the condition of that unfortunate fly of which he has spoken, the fly that suffered from a swollen abdomen." Sir Charles told the audience about a population congress he had recently attended in Berlin. "Present-day Germany must be regarded as a vast laboratory which is the scene of a gigantic eugenics experiment," he said. "It would be quite wrong and quite unscientific to decry everything which is now going on in that country. There is, as a fact, much being carried out in Germany which deserves our approbation. The authorities there are in the position of being able to carry out the advice of their scientific advisers."

So the study of genes and behavior defines the depth as well as the height of the twentieth century, because it traveled in both directions, like angels ascending and descending Jacob's ladder. The gas chambers of the Holocaust were built on Galton's principles. Auschwitz was allowed to operate to the very end of the war without Allied interference because many of Germany's enemies shared Germany's prejudices. The Holocaust was Galton's flower.

"In some sort of crude sense which no vulgarity, no humor, no over-statement can quite extinguish, the physicists have known sin," Robert Oppenheimer said after the war, speaking for the men and women he had led at Los Alamos in the Manhattan Project; "and this is a knowl-edge which they cannot lose." Many physicists in those postwar years turned from the atom to the gene as if they were turning from sin to virtue, from darkness to light. "Such a change is highly significant psy-chologically," Richard Rhodes writes in his history *The Making of the Atomic Bomb*. But biologists had lost their innocence long before the war. The study of genes and behavior was born in sin, and the possibility of sin would cling to it. After the war, the editors of *The Eugenics Review* commissioned articles about Hitler's perversion of Galton's principles ("A girl of sixteen was sterilized for answering the question 'What comes after the Third Reich?' with 'The Fourth.' "). The journal's editors were horrified, but they kept on publishing *The Eugenics Review*, with Gal-ton's flower on the cover.

BENZER WAS AWARE that a pendulum had swung back and forth between nature and nurture, propelled partly by science and partly by politics. When Galton first proposed the idea of eugenics, almost nobody understood what he was talking about. His audiences found the very word *heredity* novel and foreign. But by the time Galton published his *Natural Inheritance* in 1889 he could state as a matter of common knowl-edge that talents like "the Artistic faculty" are hereditary. "A man must be very crotchety or very ignorant, who nowadays seriously doubts the inheritance either of this or of any other faculty."

In that decade the anthropologist Franz Boas, who left Germany and physics for the United States partly to oppose the politics of Galton and his followers, began arguing the other way. Boas believed that peoples are determined more by culture than by biology. This view was advanced by Boas's students Margaret Mead and Ruth Benedict. Freud and his followers strengthened it by arguing that an individual's problems are determined more by experience than by wiring; so did the behaviorists, who argued the same way. Revulsion at the Nazi eugenics experiments made all of this intellectual current the new orthodoxy. By the 1960s, when the first molecular biologists began looking at the problem of nature and nurture, the pendulum had swung all the way back to where it had been before Galton. In 1966, when Benzer proposed his study of genes and behavior, many thoughtful people of goodwill believed that absolutely everything about a human being can be shaped from the outside. There were no innate differences between the minds of American men and women in the 1960s—or if there were, it was not politically correct to talk about them. There were no innate drives and instinctive mechanisms either; or if there were, they were not fashionable for liberal psychologists and biologists to study, not in America in the 1960s. The behaviorists' vision of human beings as blank slates was popular among liberals; they wanted the club of science to smash the hydra heads of the eugenics movement whenever those heads reared up again. The doctrine that human beings have no instincts seemed like prudent politics if not good science. In the shadow of the Holocaust the anthropologist Ashley Montagu wrote in an influential treatise against racism that a human being is "nothing but the form in which his particular culture molds his plasticity." Most American psychologists supported the learning theory, remembers Mark Konishi, a colleague of Benzer's at Caltech who now studies genes and behavior in owls: "Everything is learned, nothing is instinctive. I discuss this with Seymour from time to time, and we always laugh."

In the last decades of the twentieth century the pendulum would swing again, and this time the strongest push would come from molecular biology. Max Delbrück was one of the first molecular biologists to realize that the kind of work they did might have serious political implications. Delbrück had left the atom for the gene, but he never believed that in doing so he had escaped from sin. "Other people want to obtain power by going out into the world," he sometimes said, "but the scientist really wants to obtain power by retreating from the world." Having served his apprenticeship as a physicist in the laboratory that had discovered the possibility of atomic fission, Delbrück knew early on that a scientist can change the world more than a Hitler or a Caesar. "And you can sit very quietly in a corner and do that."

Benzer's view was narrower—or more modest. Caltech's in-house historian once asked him if the phage group had known in the late 1940s that they were the dawn of a new day. Benzer answered, "Oh, I don't know. We loved what we were doing, but I don't recall having any sense that 'we're making history.' Delbrück had a sense of history; his father was a famous historian. But my father had no history; I had no history. It wasn't part of my thinking. It was always exciting to be doing the experiments."

By the time Benzer started building his countercurrent machine at Caltech in 1966, he had seen enough of the surprises of science to know that what he and his friends were about to do might matter and that no one could predict where their work would go; although Benzer did have an idea that it might go far. Like every human being, he was really interested in human behavior. By studying genes and behavior in the fly, he was postponing what he really wanted to know, temporarily. But he suspected that if he looked deep enough, almost any creature with neurons might lead to fundamental discoveries that would illuminate every creature on the planet, including himself.

The idea of beginning the quest with flies had a kind of simplicity and whimsy that were typical of Benzer. (Feynman, the physicist, once described his own research style as "aggressive dopiness.") The project suited Benzer's down-to-earth style of talking and experimenting, and his 360-degree curiosity, the same curiosity that makes Benzer a demon to travel with, because he tries to see and explore everything, especially behind locked doors ("A closed door is always a challenge."). What Sinclair Lewis says about Martin Arrowsmith also describes Benzer: "no decorative heroisms." "He presented neither picturesque elegance nor a moral message. . . . But he had one gift: curiosity whereby he saw nothing as ordinary."

Going back to the once fabled and now scorned flies required someone with Benzer's omnivorous curiosity, his knowledge of the gene, his care and caution in the laboratory, his relaxed confidence in working at the fringes of science, and his appetite for the bizarre. Although he wore his fame lightly, the project probably needed that too: an obscure biologist could never have lured top-notch students into a project this far out. "If he hadn't done it," Crick sometimes says, "no one else would have done it."

With the distant blessings of Sturtevant and the day-by-day or night-by-night help of Ed Lewis ("I actually gave him the best technician I had," says Lewis), Benzer set up a Fly Room in Church Hall. Since he found the research interesting but could not know how the world would use it, he would, within limits, follow his curiosity. "Maybe that's a cop-out," Benzer would say to his first students in Church Hall in the 1960s. But he always ended their bull sessions and his own rare midnight soliloquies the same way: "Just do the experiments."

# CHAPTER EIGHT

---

# First Time

*—As if the idea of time had been disturbed.—*

—CHARLES DARWIN,
the M notebook

A SINGLE *DROSOPHILA* is not much bigger than an asterisk. It is so small that an escapee from a fly bottle (the flies are always escaping, always in orbit around the four corners of every Fly Room) does not even buzz—unless it flies into a drosophilist's ear. They seem silent without a microphone and they look like nothing without a microscope. But even in Benzer's first nights among his first half-pint milk bottles, he felt that fruit flies might turn out to be a magic well.

When he placed a dozen of them in an upside-down watch glass, covered the watch glass with a piece of glass for a ceiling, and observed them at low power, at twenty or thirty times life size, the flies groomed and preened, each head twisting from side to side between the forelegs, each head "all eyes." They rubbed their forelegs together in the inverted dome of the watch glass with a scheming "Ah-hah" look. When two or three met, he could see little flickering exchanges of the forelegs. The flies wandered around the watch glass, finding the very edges, and there preening again, the way a sheep will browse grass at the edge of a fence or a mouse will scratch its head at each dead end in a maze. Sometimes a fly would poke a few of its legs partway over the wall of the watch glass into the outer air and rest there, nine tenths a prisoner and one tenth free. Then, with a sudden—to human eyes instantaneous—rearrangement of parts, the fly was gone, off to another square centimeter inside the dome, exploring space and other flies.

Gradually more and more flies would cluster around the edges of the watch glass. Through the microscope the cuticles of their exoskeletons were brown but shiny, like armor. Light caught the neat red facets in the domes of their eyes. Here and there, Benzer saw a proboscis mumble its bristles against the inner walls of the glass like a baby elephant's trunk, in and out, flashing and then withdrawn. Again and again the shiny wings bent, flexing backward and downward with the movements of the hind legs. At twenty- or thirty-power magnification, every bristle on every fly's head stood out, sharp and countable—and Benzer knew that *Drosophila* had meant so much to thousands of geneticists that every bristle on the fly's head had in fact been counted and given a name.

Seeing all this action, it was easy to hope that the lessons he wanted might come out of this small theater. The flies were as quick and deft as birds and almost as expressive. When they rubbed their forelegs together, they had, to human eyes, an attitude of scheming or of prayer; and when they rubbed their wings with their hind legs, they gave an impression of agility, expertise. The neatness and deftness of their behavior matched the neatness and deftness of their bodies, both sculpted by natural selection and both intricately and invisibly linked.

A human head could grow dizzy trying to go from the flies' universe to ours, so unutterably distinct and so uncomfortably alike. Sometimes one of the flies would slip from the ceiling of the glass dome and fall flailing on its back in a panic of legs: awful to see, panic at thirty power, mildly contagious even across the gulf of the microscope.

Flies were as easy to play with as phage. "The work can be done almost pretty well anywhere," as Morgan used to say, "so long as we have a table, and an electric bulb." For Benzer, the chores were lightened because Ed Lewis already ran a spotless Fly Kitchen on another floor of the building. Working in the middle of the night, a team of lab assistants mixed fresh batches of yeast and molasses to order in a fifty-gallon vat, glopped measured amounts of fly food into the bottoms of freshly autoclaved milk bottles and test tubes, and wheeled warm, rattling racks of glassware to Lewis's and Benzer's labs every morning for their next rounds of experiments. Lewis's former technician Evelyn Eichenberger taught Benzer how to knock flies out with ether so he could examine them under the microscope without letting too many of them escape. She also set up the standard morgues in the Fly Room: traditionally beer or wine bottles with funnels in their mouths and oil at the bottom. Any flies that did escape eventually found their way down the funnels and drowned in the oil. Each morgue slowly filled with sedimentary layers of mutant flies.

Sitting in the half-dark, Benzer sent mutagenized flies by the hundreds jitterbugging through his first countercurrent machine. From almost the first runs he noticed a few individuals here and there that stood out from the crowd in one way or another. Some of the flies did not move right along—they trudged so slowly that they looked depressed. While most of the flies around them made it all the way to Tube Six, these flies moped from One to Two. Benzer collected them by sucking them up one by one with a plastic straw (guarded at the top by a piece of fine mesh) and transferring each one to a bottle of its own. When he and his technician bred those flies, many of their children and grandchildren acted the same way.

Here and there in the countercurrent machine, Benzer saw a fly go into what looked like an epileptic seizure when he rapped the machine on the benchtop. He bred those flies, and some of their children acted the same way too.

He also noticed a few flies that seemed to march right through the countercurrent machine whether the light was ahead of them or behind them. Benzer bred those flies, and once again many of their children and grandchildren acted the same way. He wondered if these flies could see the light at all, and he asked one of the first postdoctoral students in his Fly Room, Yoshiki Hotta, who had just graduated from the medical school of the University of Tokyo, to check their eyes.

After considerable work, using a microscopic electrode, Hotta managed to record signals from the tiny nerves that run between the mutant fly eye and brain. As Benzer had suspected, the electrode's readings, the electroretinograms, were abnormal. One of the first mutant flies that Hotta tested this way was *tan*, which has a light tan body and light tan antennae, and was first discovered in one of Morgan's milk bottles. The *tan* fly was half-blind.

Benzer also built a flight tester. It was a 500-milliliter graduated glass cylinder with its walls coated on the inside with paraffin oil. He and Hotta put flies in at the top. Each animal, as it fell into the cylinder, would try to fly off horizontally. Those that flew strongly would get stuck in the oil near the top of the cylinder; those that flew more weakly would get stuck lower down. Those that could not fly at all would plunk to the bottom of the cylinder. The design was pure Benzer: simple and to the point. When he and Hotta collected flies from the bottom of the flight tester they found mutants that could not fly, just as in the countercurrent machine they found mutants that could not see. Together they dissected these flightless mutants under the microscope and found congenital defects in the flies' wing muscles.

To Benzer, all these blind eyes and mangled wings were a proof of concept. But to the skeptics upstairs in the Sperry lab, they proved nothing. What did Benzer expect to find if he poisoned a fly? A sick fly. What could he learn from a sick fly? Sigmund Freud used to get the same reaction: "And you claim that you have discovered this 'common foundation' of mental life, which has been overlooked by every psychologist, from observations on *sick people*?"

Even Hotta sometimes worried that their research was way out. "When I decided to go to Seymour's lab, nobody said, 'Oh, that's a very nice idea,'" he says. His advisers at the University of Tokyo had heard of Benzer's adventures in *rII*; but not many of them liked the sound of his fly genes–and–behavior project. Packing for America, Hotta had told his professors, his friends, his family, and himself that he was willing to gamble: "'It's OK, I may be able to find something, may not.' But I didn't care. Of course," Hotta adds now, "I probably cared."

Delbrück, down in the basement of Church Hall, was feeling doubts about his own research. By now he had spent fifteen years on and off with his fungus *Phycomyces,* trying to understand something basic about the way the stalks tilt toward light—trying to see how to go from molecules to the senses. Delbrück and his students would mutagenize spores and let them grow with a light shining underneath them. Normal fungus would grow down over the rim of the agar plate toward the light. But here and there a mutant would grow straight up. Students down in the subbasement of Church Hall named the mutants that rejected the light *mad* in honor of Max.

As a laboratory organism, unfortunately, *Phycomyces* was as inconvenient as phage and flies were convenient. The fungus was harder to breed and cross, and its repertoire of behavior was, of course, limited. Delbrück was always trying to attract other fungus watchers the way he had attracted phage watchers. Sometimes he envied ethologists like Konrad Lorenz, playing outside by ponds and riverbanks. Delbrück wrote to his friend George Beadle, who was studying the genetics of another fungus, that he was "trying all kinds of things, from lunatic fringes to sober photochemistry. . . . Perhaps I should train a duck to follow me around, that sounds like a very appealing way of life."


THERE WAS NO NAME for the science they were trying to start. It was not ethology, psychology, or behaviorism. It was not classical genetics, because classical geneticists like Morgan and his Raiders or Ed Lewis arranged their crosses and mapped their genes without reference to molecules. Nor was it behavior genetics, because behavior geneticists also bred animals without studying the underlying molecules. It was not traditional neurobiology, the study of the workings of nerves and brains; Sperry and his students were neurobiologists, and they were less than impressed. Benzer did not think that what he was doing was molecular biology, either, because his prime interest was animal behavior. An atomic theory of behavior was new science. Since it is a research program that moves from gene to nerve and from nerve to behavior, it is sometimes called neurogenetics. Given its deep roots in the natural sciences, it might also be called natural psychology. "I don't care what you call it," Benzer says. "I've often said, I don't care about disciplines, I care about nondisciplines. What do you care about names?"

In 1967 Benzer published his first paper on the new work, "Behavioral Mutants of *Drosophila* Isolated by Countercurrent Distribution," a paper that scientists in several disciplines now think of as a landmark. He was forty-six. He left Purdue, joined the biology department at Caltech, and moved into the laboratory space in Church Hall where he still works today.

Benzer had written that first paper as a kind of mission statement for his research program. But the project that proved the plan could actually work was done by a graduate student of Benzer's, Ronald J. Konopka, from Dayton, Ohio. Konopka joined the laboratory because he wanted to use Benzer's genetic scalpel to find and dissect the sense of time. He thought there must be a master clock hidden in the clockwork of life, and he thought Benzer's method was the way to find it. Morning glories know when to bloom. Bears in caves know when to wake up. Grunions know when to spawn. ("And doubtless, when swallows come in the spring, they act like clocks," says Descartes.) During the Enlightenment the French astronomer Jean-Jacques d'Ortous de Mairan performed a famous experiment with heliotrope, a plant whose Latin name means "turning toward the sun." The heliotrope's leaves and stems unfold every morning when the sun comes up and fold again every evening when the sun goes down. In the summer of 1729, the astronomer dug up a single heliotrope plant, brought it inside, put it in a pitch-black room, and peeked in every now and then. Even in the dark room the plant was raising its arms and lowering them again, keeping time with the heliotrope out in the garden. A note about this experiment appeared that year in the *Histoire de l'Académie Royale des Sciences:* "The sensitive plant follows the sun without being exposed to it in any way. This is reminiscent of that delicate perception by which invalids in their beds can tell the difference between day and night."

The astronomer's experiment inspired innumerable imitations. A French botanist carried "sensitive plants" down into a wine cave and watched them by candlelight—not a bad research project. A Swiss botanist tried growing sensitive plants in a room lit only by banks of lamps, and he managed to change the plants' behavior by lighting or snuffing the lamps. The great Swedish naturalist Carolus Linnaeus dreamed of putting together a flower clock made of evening primrose, marigolds, childing pink, scarlet pimpernel, hawkbit, bindweed, nipple wort, passion flower, spotted cat's ear, and the Star of Bethlehem, "by which," Linnaeus wrote, "one could tell time, even in cloudy weather, as accurately as by a watch." The passion flower would open at noon, the evening primrose at 6 p.m., and so on. One summer, Darwin drew a series of diagrams of a single leaf of Virginia tobacco as it stirred upward and downward from three in the afternoon to 8:10 the next morning.

Silverfish, crickets, spiders, scorpions, and squirrel monkeys have a sense of time. Biologists proved this by simple experiments, building exercise wheels like the wheel in a mouse cage and monitoring an animal's cycles of sleeping and waking on the wheel in windowless rooms that were never dark or never light. Twentieth-century biologists built exercise wheels like these for sea hares, lizards, and cockroaches; and they built balances like seesaws so that every time the creature moved, it tilted the seesaw. The more they looked, the more they discovered that the sense of time is everywhere. Even single-celled animals such as *Euglena* have a sense of time. *Euglena* swims like an animal but has green chlorophyll like a plant. In a pond, each cell swims more by day than by night; at night it tends to sink sluggishly downward and downward through the water column. And even in a lab in constant light it keeps to this rhythmic cycle, which biologists call "circadian," meaning "about a day."

The literature on circadian rhythms is full of strange factoids about the sense of time in living things. The cells of a banana divide just after dawn. Some animals have clocks longer than twenty-four hours: a normal human clock, for example, runs a bit slow in a cave or a windowless room; we automatically adjust it every day to keep in time with the sun. Our clocks are reset by the dawn. The clocks of mice run a bit fast, and they are reset by the dusk. Confuse the sense of time of a homing pigeon—shift its clock six hours by tricking it with lights—and it will make a ninety-degree error in the path of its flight. Many people can order themselves to wake up at, say, seven in the morning—and they will wake up within a few minutes of seven. "All this showing," as Spinoza says, "that the body itself can do many things from the laws of its own nature alone at which the mind belonging to that body is amazed."

By the middle of the century most biologists assumed that this sense of time must be in the genes. But contrarians argued for nurture, not nature. Living things might learn to keep time with the beat of the sun the way goslings learn to follow Mother Goose or Konrad Lorenz. They argued that each seedling, gosling, and newborn baby might be imprinted by the sun, learning the rhythm of day and night and then keeping the beat, keeping time with the sun for the rest of their lives. Some biologists speculated that even sensitive plants sequestered in a wine cave might keep time by picking up subtle tides in atmospheric electricity or tides of cosmic rays, or perhaps secret signals linked to the phases of the moon, cycles of sunspots, or the rotation of the planet.

To prove that the sense of time is a matter of nurture, not nature, a biologist at Northwestern University, Frank A. Brown Jr., designed elaborate experiments involving carrots, seaweed, crabs, rats, and much solitude and darkness. He grew potato plugs in sealed jars and monitored the rhythms of their metabolism by sampling carbon dioxide and oxygen with gas detectors. He shipped oysters from New Haven, Connecticut, to his laboratory in Evanston, Illinois, and watched them open and close their shells rhythmically day after day in pans of seawater. In another herculean experiment he watched and timed 33,000 individual mud snails as they crawled out of holes. He also lobbied the U.S. National Aeronautics and Space Administration to put one of his potatoes into orbit. He wanted to see what would happen to its metabolic rhythms when the potato was aloft in a satellite and could no longer feel the rotation of the planet. NASA never took up his potatoes.

In 1960, a botanist tested Brown's hypotheses in a cheaper experiment by flying Syrian golden hamsters, among other creatures, to the South Pole. There he put the hamsters and their cages on a turntable so that their rotation would counteract the rotation of the Earth. (The Earth's rotation cannot be counteracted this way except at the pole.) The hamsters went right on waking and sleeping at the same times and with the same rhythms as hamsters that were not revolving on turntables. Apparently the hamsters did not need any cues from the rotation of the Earth to keep track of time. The botanist also put bean plants, fungi, cockroaches, and fruit flies on his turntables. All of them kept time perfectly.

After that experiment, it seemed clear to almost every biologist on the planet (except Brown) that living things really are born with some kind of inner clock. But no one knew where the clock might be hidden in the body or how it might work. One test of a true clock is its ability to keep time through a wide range of temperatures. A clock that speeds up in hot weather and slows down in cold weather is not a clock, although it may make a good thermometer. So a drosophilist checked the clock in fruit flies by raising them at different temperatures. Heat and cold did nothing to change their rhythms. Whatever they had inside them, wherever it was concealed, and however it worked, it really did deserve to be called a clock.

In 1969, at a meeting on the sense of time, the botanist who had gone to the South Pole, Karl Hamner of UCLA, declared that the problem was as mysterious as gravity before Newton: "What we need now is another Newton."

MEANWHILE, Konopka was experimenting in Benzer's laboratory. The behavior the flies are named for is waking up in the morning: *Drosophila* means "lover of dew." In fact, the flies display their love of dew from the moment they are born. Each young fly develops inside a pupal case. When it is ready to emerge, the fly does not pip its shell with a beak, like a bird; instead it inflates a tiny balloon on its head, like a steering-wheel air bag, and bursts right out of the pupa. Benzer thinks a fly emerging from a pupa is one of the sweetest things in nature. The fly crawls out wet, like a newborn baby. Its wings are not yet inflated; they are crumpled like a new butterfly's, and because of the air bag its head is still disproportionately large, again like a newborn baby's. In nature, the flies usually emerge around dawn, when the world is moist and dewy. Even if a jar full of fly pupae is taken out of the light and placed in total darkness for several days, the young flies will still emerge together in the dark, around the time of their virtual dawn.

By the time Konopka went to work with Benzer, biologists had already mounted round-the-clock watches on fruit flies. They had discovered that normal flies live on a daily cycle just like the astronomer's heliotrope. At sundown a normal fly becomes very quiet. It does not close its eyes, but it stops moving and looks as if it is asleep on its feet. At sunup it begins to move around again. Flies will keep to this daily cycle even in total darkness, just like the heliotrope.

In Benzer's laboratory, Konopka poisoned flies with EMS to make random "typos" in their DNA. He used these poisoned flies to establish hundreds of separate lines of mutant flies. When he was finished, he had hundreds of fly bottles, and each bottle contained a separate line of mutants. Konopka spent most of the summer of '68 watching these mutants' children eclose in bottle after bottle, looking for flies that missed the dawn. It was a lonely way for a young man in California to spend the summer, and most of the professors and students who passed by his door thought he was wasting his time. The smart money said that Konopka had chosen a piece of behavior too central and too complicated to dissect through the genes. A mechanical clock has hundreds of gears, springs, screws, and ratchets. A living clock might require hundreds of working parts too, and hundreds of genes to make each part. But the parts of a clock are complexly interdependent, and if EMS caused a mutation in any one of those hundreds or thousands of genes, the result for the fly was likely to be a broken clock. That is to say, any of hundreds or thousands of mutations would have the identical effect on the fly, destroying its sense of time. So even if Konopka did find a clock mutant, he still might not have a clue what had gone wrong inside it. A fly without a clock might not even live long enough to eclose.

The ragging that Konopka endured that summer would later become a legend in the Benzer lab and far outside it. Geneticists and molecular biologists would tell the story to the tune of "They Laughed at Columbus." "They said, 'It's too much work!' " says Jeff Hall, who was another of Benzer's first postdoctoral students. "They said, 'You'll never find them!' They said, 'If you make one, it will die!' But he told them, 'Bugger off!' "

Konopka kept all of his flies at a constant temperature in cycles of twelve hours of white fluorescent light and twelve hours of darkness. For the sake of sanity and simplicity, he checked the bottles only twice a day: just after the lights went on in the morning, and just before the lights went off in the evening. If the flies had a normal sense of time, very few of them would eclose before that first check in the morning. When he inspected their bottle in the morning, it should still be full of eggs. The flies would eclose from those eggs in the next few hours of light, and he would find them creeping, crawling, and flitting around in the bottle when he checked on them that evening. So if Konopka checked one of his fly bottles first thing in the morning and found dozens of newborn flies inside it, he would know that they had eclosed sometime in the night, and he would suspect that there was something wrong with their sense of time.

In bottle after bottle, the flies behaved normally. But in the two hundredth bottle, when Konopka checked in the morning, he saw that it was teeming with flies. And when he inspected them one by one, he saw that most of them were males. Konopka bred those males. When it was almost time for their children to eclose, he put them into absolute darkness and waited to find out what they would do. Not many eclosed at dawn. They eclosed at all hours of the day and night, just like their fathers. He had found his first clock mutant.

In a second bottle, Konopka discovered a line of mutants that eclosed at the wrong time too. And when he mounted a careful watch over them he saw that they eclosed too early. Apparently their dawn came sooner than it came for the rest of the world. And in a third bottle he found a line of mutants that eclosed too late.

Konopka talked over the case of these three mutants during a long, meandering lunch with Benzer and Hotta and a few other students. They all wondered what the mutants' sense of time would be like after they eclosed. Over lunch, Benzer thought of a quick way to help Konopka find out. As Benzer tells it now, this was another small countercurrent moment. "The people in the lab were split. One guy said, 'That'll never work. If that works, I'll buy you an Indonesian dinner.'"

So Benzer tried it. He went back to a standard gadget from physics and chemistry, a spectrophotometer. The centerpiece of a spectrophotometer is a little glass square-sided cylinder called a cuvette. When physicists or chemists have a mystery substance to identify, they pour a few drops of it into the cuvette and switch on the spectrophotometer. The gadget fires a series of light beams through the cuvette: all the colors of the rainbow, plus ultraviolet and infrared. A sensor analyzes each beam as it passes through the cuvette, and a marking pen makes a series of squiggles on a rolling drum of paper. Sometimes the investigators can identify their mystery substance from the pattern of the squiggles on the paper.

Benzer put two strips of black tape on the outside of the cuvette, with a little space between them for the beam to pass through. Then he put one of Konopka's flies inside the cuvette, and corked it with a cotton plug. He set the spectrophotometer to infrared, which flies can't see. Whenever the fly moved from one end of the cuvette to the other it would pass between the two strips of tape and block the infrared beam, and that would make the pen jiggle up and down on the drum of paper. When he was finished, Benzer turned on the machine and let it run all night.

The next morning, Konopka arrived at the second floor of the Church Laboratory and found paper spewed out all over his floor in an almost endless scroll. He fished through loop after loop of paper and looked at the bursts of inky squiggles. The trick worked. He could see exactly how busy the flies had been every minute of the night.

("So that was a very good dinner, actually," Benzer says. "J.J.'s Little Bali, near the airport, in Inglewood.")

Later, Benzer's postdoc Yoshiki Hotta built a whole set of gadgets based on the same principle, so that Konopka could monitor many flies at once. Now Konopka could find out what his mutants were doing in the dark. A few days of monitoring told him that his first line of mutants was consistent in its inconsistency. Not only did these mutants eclose at all hours of the day and night; for the rest of their lives they woke and slept, wandered and paused at all hours of the day and night. They acted like insomniacs. They seemed to be time-blind.

The line of mutants that eclosed early was also consistent. The day after they eclosed, they woke up about five hours too early; they did the same thing for the rest of their lives. Apparently these mutants did have clocks, but the clocks ran too fast. Their days had a period of nineteen hours. And Konopka's third line of mutants, the line that eclosed late, was just as consistent. They woke up late every day of their lives. Their clocks ran slow. Their days had a period of twenty-nine hours.

Konopka examined the mutants through a microscope. All of them, females and males, throughout all the stages of their life cycles, from egg to larva to pupa to adult, looked absolutely normal. And when they bred, they passed on their sense of time from generation to generation. He could see that in bottle after bottle only the mutant flies had a warped sense of time.

If Konopka had been working with chicks, chameleons, monkeys, guinea pigs, potatoes, or heliotrope, he might not have been able to push this work any further. But with *Drosophila* the next step was obvious. He crossed his short-period mutants with a few classic mutants from the Fly Room, including *white, singed, yellow,* and *miniature.* Methodically, using cross after cross, he began trying to map the short-period muta-tion, using the same method that Sturtevant had invented in Morgan's first Fly Room. The method was the same, but now he was trying to map a gene that was manifest not in the color of the fly's eyes or the shape of its wings but in its behavior: a mutation that changed the way it moved through time.

Konopka found that the short-period mutation mapped to the far-left end of the X chromosome, less than one map unit from *white.* When he mapped the arrythmic mutation, he found that it too was on the far-left end of the X chromosome, also next to *white*. He mapped the long-period fly, and again it was on the far-left end of the X chromosome, also next to *white*.

By now geneticists called the map unit the *centimorgan,* in honor of the man who had started their science. If there is a 1 percent chance that two genes will be separated by crossing-over, those two genes are said to be separated by one centimorgan. Konopka's three mutants were less than one centimorgan away from *white,* and they were zero centimorgans apart from each other.

Now Konopka was amazed. These were the first three time mutants he had found, and they all mapped to precisely the same place. Because they mapped to the same place, they had to be alleles, or variants, of the same gene, like tall and short in Mendel's peas. Konopka had looked at more than two hundred strains of mutants to find these three time mutants, and all three of them pointed to the same spot on the same chromosome. He had barely started his search, and already he seemed to have stumbled straight into the center of the living clock.

By arranging marriages for his mutants, Konopka created flies that had one normal copy of the clock gene and one mutant copy. He also made flies that had two normal copies and flies that had two abnormal copies. It was just like breeding Mendel's pea plants—tall-short, tall-tall, short-short—but this was behavior. He monitored the flies' children and grandchildren, reading the scrolls of paper day after day. He could see that two of these mutations were at least partially recessive, like short-ness in peas, white eyes in flies, or blue eyes in human beings. The short-period mutation was partially recessive. The mutation that destroyed a fly's sense of time was also recessive. That is, if a fly inherited one broken copy of the gene and one normal copy, its sense of time was almost nor-mal—its clock ran just half an hour slow.

In test after test, all three mutations mapped to the same spot. They were clearly alternative versions of the same gene. In the jargon of genet-ics, each spot on a chromosome is a locus. Konopka had discovered three alleles of a locus on the X chromosome that shapes the fly's sense of time. He had now earned the right to name the gene. Because a change in the gene had the power to change the period of the fly's days, Konopka called it the *period* locus.

He had found a very peculiar gene, and he had found it in his first two hundred bottles. Later on, when Benzer and his students, building on this first success, began to wander in many new directions, Konopka would formulate Konopka's Law. It was his first law, and so far it is his only law: "If you don't find it in the first two hundred, quit."

WITH THE DISCOVERY of the clock gene, the sense of time, mysterious for so many centuries, was no longer a mystery that could be observed only from the outside. Now it could be explored as a mechanism from the inside. The discovery implied that behavior itself could now be charted and mapped as precisely as any other aspect of inheritance. Qualities that people had always thought of as somehow floating above the body, apart from the body, as if they belonged to the realm of the spirit and not of the flesh, as if they were supernatural, might be mapped right alongside qualities as mundane as eye pigment.

At the time, not many people at Caltech or elsewhere were prepared to believe Konopka's mutants or his maps. They could not believe that his X marked the spot. "People really resisted the notion that this had anything to do with the phenomenon," Konopka says now. "They could never get it through their heads what it meant that these mutations were all at the *same locus*." Three different mutations in one locus meant that he had found not just a piece of the clock but a central piece, maybe *the* central piece, the pacemaker of the fly's behavior, the piece of living machinery that keeps it waking and sleeping and moving in time with its planet from the moment it is born. His map suggested that a single gene can shape vast arrays of behavior, that individual genes can have extraordinary power to influence a life. He could even hope (though it was only a hope) that the gene he had found in the fly would tell him something about the mechanism that drives our own human sense of time.

As Konopka made more and more crosses and his map became more and more convincing, Benzer and Konopka got excited—but not the skeptics up and down the hall. "They would try to deny it," Konopka says now. "They couldn't think about it. They didn't appreciate the power of genetics. They refused to believe that anyone would get the pacemaker." Ever since the turn of the twentieth century, biologists had been trying to approach the mysterious center of life by way of genes. "But they would refuse to believe someone would have a handle on this."

Not even Hotta trusted Konopka's results. "I cooperated with him very heavily," Hotta says now, "and I constructed the machines he used to assay the behavior. But at that time, I was rather skeptical about the gene, and so I dared not put my name in the paper." Konopka and Benzer

*"I don't believe a word of it!" Max Delbrück doubted the genes-and-behavior stories that began to pour out of the Benzer lab in the second half of the 1960s. Here, Seymour explains and Max doubts, after a seminar at Caltech.*

wrote up a report, "Clock Mutants of *Drosophila melanogaster*," and sent it to the *Proceedings of the National Academy of Sciences.* At a party in Pasadena, Benzer told Delbrück that they had found alleles of a new gene linked to behavior. Benzer explained why he and Konopka thought the mutants had something wrong with their sense of time.

"I don't believe it," said Delbrück.

This scene also became part of the Konopka legend. Konopka was standing right next to Benzer and Delbrück at the time.

"But Max," Benzer said, "we found the gene!"

"I don't believe a word of it," said Max.

# CHAPTER NINE

# First Love

*What is it men in women do require?*
*The lineaments of Gratified Desire.*
*What is it women do in men require?*
*The lineaments of Gratified Desire.*

—WILLIAM BLAKE,
*"The Question Answer'd"*

DARWIN DIVIDED the adaptations of life into two kinds: the ones we need to survive and the ones we need to reproduce. The clock is one of the oldest adaptations of the first kind. Living things needed clocks as soon as they had begun to accumulate other adaptations: they needed clocks to organize the rest. Having a clock allowed the first simple life-forms billions of years ago to grow on a schedule—to make any compounds they required for photosynthesis before sunup, for instance, and to taper off their production before sundown, as plants still do today; or to hunt for other living things to eat when those things were most plentiful and most vulnerable—which owls and wolves still do today, each to its own clock. Inventing a clock was probably one of the first acts of life, and that is why clocks are ubiquitous. With *period*, Benzer and Konopka were looking at the first known specimen of one of the oldest instincts on the planet.

What is going on in our heads when we feel the passage of time is a question that they were wise enough to ignore, temporarily. That problem has defeated philosophers for millennia. Bishop Berkeley tried to define time and found himself "embrangled in inextricable difficulties." Saint Augustine said that we all know what time is until we try to put it into words. The Roman philosopher Plotinus thought the sources of time must lie inside us, that time springs from the human soul.

A clock gene is not the same as the sensation of time, any more than a cascade of molecules in the retina is the same as the sensation of sight. Still, without rhodopsin and a long string of other molecules we are color-blind, and without clock genes we are time-blind. So *period* is a way into one of the sources and wellsprings that Plotinus believed must flow from the soul in "all the dense fullness of its possessions."

There was something thrilling but also peremptory, down to earth, even absurd about the discovery of the first clock gene, as there would be about all the rest of the discoveries that followed. To go from the contemplation of time to the contemplation of clock genes means coming down to earth with a bump. To turn from the sublime to a mechanism so anatomically concrete can make us feel ridiculous. In his celebrated M notebook (M for Metaphysics, Materialism, and Mind) Darwin jotted a note to himself about Plato. "Plato says that our ideas arise from the preexistence of the soul, and are not derivable from experience," Darwin wrote, and he added "—read monkeys for preexistence."

But there are wheels within wheels even in *period.* Eventually molecular biologists in laboratories around the world would be bent over Konopka's clock gene like scholars bent over a single verse of Hebrew or Greek in which they could almost read the secret of life. "Parmenides," writes David Park in his book *The Image of Eternity: Roots of Time in the Physical World,* "was known principally for a poem, written in high poetic style, in which he analyzed the mysteries inherent in the single Greek word *esti,* 'is.' "

TO TURN FROM the springs of time to the springs of love means coming down even harder. Love is what wise men and proverbs do not pretend to explain. "Three things are too wonderful for me," a verse declares in Hebrew Scripture, "four I do not understand":

> *the way of an eagle in the sky,*
> *the way of a serpent on a rock,*
> *the way of a ship on the high seas,*
> *and the way of a man with a maiden.*

In *The Gold Bug Variations,* a novel about the cracking of the genetic code, Richard Powers imagines a meeting between Albert Einstein and

T. H. Morgan at Caltech. Morgan explains what he is trying to do in his Fly Room, the union he is trying to arrange between biology, chemistry, and physics. "No, this trick won't work," Einstein cries. "How on earth are you ever going to explain in terms of chemistry and physics so important a biological phenomenon as first love?"

In Darwin's terms, the adaptations of reproduction are as old as the adaptations of survival. Reproduction is one of the defining acts of life; and without reproduction there would be no way for the Darwinian process to begin, since Darwin's process is evolution by the selective success and failure of populations of reproducing forms. Small differences written in single changes in the letters of the double helix led rapidly under the pressure of natural selection to an extraordinary profusion of forms and also to forms of self-advertising, courtship, and copulation that are as miraculous as any phenomena in the natural world.

If the clockwork gene stands for all the clockwork of the body's apparatus of survival, then the instincts of reproduction stand in our minds for all the miraculous complexity of behavior. Males and females need displays to find each other, to recognize each other, and also to impress each other, since virtually every copulation in the world takes place beside a big gene pool of competition. This produces powerful evolutionary pressures that Darwin called sexual as opposed to natural selection.

Galton assumed that other animals cannot vary their courtship routines the way we humans do with fashions. But humpback whales sing songs that radiate through the oceans for thousands of miles and change from season to season much like the Top Ten songs that fill the airwaves over the whales' heads. At any one moment in any one ocean all of the males sing the same song. But within a month they will all be singing a new song, and unlike human beings they never sing a golden oldie from a decade or two back. They never repeat themselves. The songs appear to be courtship songs, elaborate displays, like most of ours on the radio. They actually include the use of rhyme; and according to Roger Payne, who has been recording them since the 1960s, the love songs of some whales are audible more than ten thousand miles away. "When you swim up next to a singing whale through the cool blue water," Payne writes, "the song is so loud, so thundering in your chest and head, you feel as if someone is pressing you to a wall with their open palms, shaking you till your teeth rattle."

Male bowerbirds in Australia and New Guinea do not sing fancy songs or flash fancy plumage. Instead they build bowers, pretty little shelters, each species according to its own design. Some build teepee style, with branches leaning in against a sapling that bowerbird watchers call a "maypole." Others build what are known as avenue bowers, which they invite females to walk through. The satin bowerbird paints the walls with a twig brush; he makes the paint out of chewed fruit, charcoal, and his own saliva. Other bowerbirds drag in live orchids. They throw out the old, wilted flowers every day and redecorate the bower with fresh flowers. Males trash each others' bowers, steal each others' flowers, and even barge in sometimes to interrupt other pairs in coitus. The satin bowerbird, which has bright blue eyes, decorates its painted bower with anything blue it can find, according to the ornithologist Frank Gill: "One bower was decorated with parrot feathers, flowers, glass fragments, patterned crockery, rags, rubber, paper, bus tickets, candy wrappers, fragments of a blue piano castor [*sic*], a child's blue mug, a toothbrush, hair ribbons, a blue-bordered handkerchief, and blue bags from domestic laundries."

These spectacular animals would have been impractical subjects for attempts to isolate links between genes and behavior. Again the molecular biologists had to start simpler. Sydney Brenner studied mutants of courtship and copulation in the nematode worm. The worm is slow and undulant. Watching it through a microscope at high power is like watching whale sinuosities through a porthole. It glides around a petri dish on a bed of agar and around its food, a blob of *E. coli* in the center of the agar. Generations in Brenner's school have now gotten to know the slither look of them, the male nosing around the female—wrapping around her, searching efficiently for the vulva, and finding it with the ingenuous directness or excitement of the young Philip Roth. Brenner and his students learned the worm's ways and habits, which are highly regular ("Let's give it forty-five seconds, and it will defecate again."), and they learned to pick them up with toothpicks or titanium wire and to freeze and unfreeze mutants for their genetic dissection experiments. Like the drosophilists, they fell in love with their animals and saw more and more richness of behavior. There is a worm lab just down the hall from Benzer's Fly Room at Caltech. It is run by a young molecular biologist named Paul Sternberg, who keeps three or four thousand strains of mutants, double mutants, and triple mutants frozen in liquid nitrogen for breeding experiments. Often when he and his group make mutants and something goes wrong, he becomes aware for the first time of some piece of normal behavior that he hadn't noticed before. "If you're narrow-minded like we are," says Sternberg, "you don't realize it until you perturb it. Then you go back and see it. That's the code of the geneticist. Of course, an animal behaviorist, an ethologist, would just watch. But I hang around with people who get excited by genes. They get excited when they can talk about genes."

In the Hawaiian archipelago, sexual selection pressures have produced fantastic displays even among fruit flies. There are more than four hundred species of *Drosophila* in the islands, and they all descend from a few flies that blew in on a freak wind millions of years ago, possibly from a single pregnant Eve. Being lovers of dew, the fruit flies live mostly in the rain forests on the cool green windward sides of the volcanoes. The picture-winged *Drosophila* are some of the most striking-looking—big for fruit flies, six to eight millimeters long—and their courtship is striking too. Courting males fly to a solitary spot on a tree trunk or a large leaf or fern a few feet above the ground. As many as ten males, sometimes from three or four different species, stake out different fronds of the fern or scales of bark or petals of an orchid. This kind of courtship is known as "lekking": male fruit flies' equivalent of hanging out at the 7-Eleven. They do wait around almost from seven to eleven—from sunrise to sunset, even through light rains. Males of some species stay motionless; others perfume their spots with tiny anal droplets of male pheromones as advertisements for themselves. Female picture wings are not ready to mate until they are a month old, and they often live another month after that, and they mate only once. So each female takes her time flying from lek to lek, day after day, sometimes for weeks, before she settles down to court and be courted.

Close up, males of each species court in their own way, according to the drosophilist Herman T. Spieth. A male *Drosophila ornata* stands directly behind the female with his head under her wing vanes, extends his proboscis, alternately stamps his forelegs against the fern, spreads his wings, straightens his abdomen, elevates the tip, and pulsates an anal droplet. After that, he goes into routines and subroutines. For instance, if the female kicks rearward with her hind legs, Spieth says, the male typically backs away several millimeters, spreads his wings horizontally to forty-five degrees, and exposes certain selected segments and membranes. The male sometimes courts from in front, and there his routine is completely different, including many ritual pawings of the fern, curlings of the abdomen, and what Spieth describes clinically as "contact of the labellar lobes," explaining, in parentheses, "(kisses)."

*Drosophila hamifera* has a different display, including wing vibrations and Elvis-like gyrations of the abdomen. "If the female responds by making labellar contact (kissing), the male then circles rapidly to her rear."

There he assumes their species's ritual position, says Spieth. He puts his head under her wings, holds on to her with his hypertrophied labellum (swollen lips), thrusts his forelegs under her abdomen, and moves them alternately to and fro. . . .

And so on: innumerable distinct routines and if-then subroutines in a single group of flies on a single isolated group of islands. Many of these pieces of behavior may have evolved the way the flies in Benzer's Fly Room evolved, bit by bit, with changes of one letter or a few letters of genetic code at a time. The routines help the males do what the humpback's songs and the bowerbirds' bowers do: advertise a male as a good choice and set him apart from the rest of the lek. And a surprising number of the Hawaiian fruit flies' routines include a moment in which the female, after a period of ritualized dodging, stabbing, darting, stamping, fanning, or apparent indifference, will march forward head to head "with the labellar lobes open and firmly 'kiss' the male's open labellar lobes."

After the female copulates with the male, she takes off and never sees him again.


"SURPRISINGLY," says Michael Ashburner, who is a world authority on *Drosophila* at Cambridge University, "most entomologists don't regard *Drosophila melanogaster* as an insect." Ashburner laughs. "Well, it is, but because the literature on *Drosophila* is so huge and a lot of it requires you have some understanding of formal genetics and most insect biologists don't, they're scared of it, basically."

Nevertheless, *Drosophila melanogaster* is made of the same stuff as other insects. Put a virgin female and a male *Drosophila melanogaster* beneath a large watch glass, and the action runs much the same course again and again, more or less like clockwork. The male sees the female, and even if he has never seen a female fly before in his life—even if he has never seen another living thing in his life—he seems to experience, as Benzer puts it, "an immediate 'aha!' reaction," a fly Adam noticing Eve. Within seconds he maneuvers so that he is facing her head from one side. Then he holds out one wing toward her in a kind of salute and sets it vibrating: the love song of the fly. Hurrying around to her other side, he holds out his other wing toward her and vibrates that: second verse, same as the first. To Benzer in his workroom in the middle of the night, the song was just barely audible if he lowered his ear to an open vial: *Eine Kleine Nachtmusik*. The song of the fly does not sound romantic to a human ear, but then the fly seems unmoved by human songs that blast from the radios, tape decks, and CD players in the Fly Rooms. When Benzer played the fly's love song at lectures, he used to introduce it with P. T. Barnum–style blarney: "You are privileged to hear a recording of the male courtship song of *Drosophila melanogaster,* and I hope the ladies will control themselves." He enjoyed pointing out that at the female fly's antenna, which is a sound-receptive organ, the love song has a volume of about one hundred decibels, which is comparable to the climax of the "1812 Overture."

The male fly does not just shake its wing at the female at random. *Drosophila melanogaster* is thought to have evolved in Africa and must have had to set itself apart in those rain forests just as much as drosophilids did in Hawaii. If a male *melanogaster* does not sing just the right song, a female will emit a counterbuzz, a rejection buzz, which is an international fruit-fly message that males of every species seem to understand. Sometimes she will bat him away during his performance or extrude her ovipositor in his face, a sight that seems to have a discouraging effect. But if the male is singing her song and if she is a virgin, the rest of the action proceeds in the distinctive *melanogaster* series of steps, which are, as Benzer says, "only too embarrassingly anthropomorphic." The male has an erectile penis. The female has a vagina. Copulation typically lasts twenty minutes. ("How anthropomorphic can you get?")

Courtship and copulation are behavior of a higher order than the kind of behavior that is driven by clock genes. Courtship requires a whole series of complicated steps, a long chain of different pieces of behavior, and each step makes the next step more likely: One thing leads to another, as we say. Flies inherit every step in this dance. When a male courts a female, first he taps her with his foreleg, as if to get her attention. Then he follows her around, and starts singing. After the song, he sticks out his proboscis, as if to ask, Do I really want this? Is this female? The right species? He kisses her, licks her, and finally attempts to copulate. Richard Feynman invented a way of drawing the interactions of subatomic particles with arrows to show them approaching each other, and more arrows to show the parting of the ways. Back in the days when Benzer still thought of flies as simple particles of behavior, he sometimes drew the flies' sequence of courtship steps in the style of a Feynman diagram with a helix in the middle. It was a joke in the same spirit as his countercurrent machine—describing living things as particles of behavior—but of course courtship and copulation are behavior of a much higher order than the behavior of particles. And again, the fly inherits all of this behavior along with its body. Behavior is part of the package.

Benzer could not imagine an instinct more interesting to investigate, and he thought that with genetic dissection he might be able to tease apart the steps. "The trick is designing a screen," explains Ashburner of Cambridge. Benzer's countercurrent machine was an ideal screen for phototaxis, Ashburner says ("Very, very simple but very elegant."). His flight tester was an ideal screen for flight mutants ("Again simple but elegant."). Screening for time-blind mutants was a little harder, because one had to figure out a way of checking thousands of flies for a fly that had a weird sense of time. "But Ron Konopka did that," says Ashburner. "And damn painful some screens can be! And there are still aspects of complex behavior which are extremely difficult to—to—to—I mean, just logistically difficult to get mutants in. Certain aspects of sexual behavior, where you have the males posturing to the females. Or licking her bum, or whatever. You know. You could imagine you could do it, but I mean in technical terms it would be logistically an horrendous task to—"

In 1971, the year Benzer and Konopka published their discovery of clock genes, Jeff Hall joined the laboratory as a postdoc and began trying to figure out how to screen for mutants of courtship and copulation. Of all Benzer's students, Hall was the one with the deepest background in *Drosophila*. He had been working since his undergraduate days with fruit flies, fly bottles, and funnel-in-beer-bottle morgues. Hall and Benzer decided that the simplest way to go into the problem would be to screen for what Hall was soon calling, in an ironic and rueful tone he learned from Woody Allen, *savoir-faire* mutants, flies that have no luck in love. To find them, Hall borrowed a set of mutants from another fly laboratory—a line of mutants in which the males never fathered any children. Hall put these males together with normal females and watched to see what would happen. The males could sing the love song of the fruit fly and they could follow it up—they could talk the talk and walk the walk—but they were sterile. They had defects in their reproductive systems.

Those were not very interesting mutants. They were as boring as flies with blind eyes or bad wing muscles. On the other hand, Hall could not help feeling impressed by the power of the instinct these flies had inherited. He had assumed that a blind fly, for instance, would be a *savoir-faire* mutant. But in a fly bottle a blind virgin male raised in total isolation can find a virgin female, even if she is blind too. Apparently he sniffs out her aphrodisiacal advertisements, her pheromones. The two mutants meet and pass on their genes.

Hall had also assumed that if a male could not fly he would not be able to hold out his wings and make the fly's tremble song. But even the flightless mutants that Benzer found in the bottom of the flight tester were willing and able to sing a love song down there. When these flightless mutants spy females, Benzer says, they vibrate their wings "in a quite normal way, yet, when dropped off the end of a rod into open space they just clunk right down to the top of the table." Once, years later, a student in Hall's laboratory made male double mutants that could neither see nor smell. Then he cut off their wings. He introduced them to female double mutants that could neither see nor smell. A few of the double mutants still mated.

EACH TUFT of grass brings forth seed "after his kind," as it is written on the first page of the Book of Genesis. Every life-form brings forth "after his kind," the great whales in the waters and the birds above the waters. And every living thing that swims, creeps, or flies bequeaths to the next generation not only the form of its kind but also the instincts of its kind, including the instinct of generation itself: "And God blessed them, saying, Be fruitful and multiply, and fill the waters in the seas, and let fowl multiply in the earth." This has always been one of the primary miracles, the burden of the first page.

Benzer and his students were trying to find a point of entry into these instincts by looking for places where the instincts had gone off on a new bent. Sometimes they noticed mutant flies that were half male and half female. Years before, Morgan's Raiders had noticed these flies too. They are called "gynandromorphs," from the Greek gynē, meaning "woman," and andr, "man": morphs of male and female. In some gynandromorphs the right half of the body is male and the left half is female. Every cell in the male half has one X chromosome and every cell in the female half has two. In some gynandromorphs—gynanders for short—the split runs down the middle, passing right through the head: the right eye is male, the left eye is female. In other gynanders the sexual border cuts across the body on a diagonal. The female melanogaster is bigger than the male, so the female parts of the gynandromorph are bigger than the male parts. Also, a female melanogaster's abdomen is brown; a male's is black (melanogaster means "black belly"). So the gynander's abdomen is a motley brown and black. "Glory be to God for dappled things," the poet Gerard Manley Hopkins wrote, praising "skies of couple-color as a brinded cow," "landscapes plotted and pieced," "all things counter, original, spare, strange." Of all things counter, original, spare, and strange, the gynander is one of the strangest—maybe too strange for Hopkins.

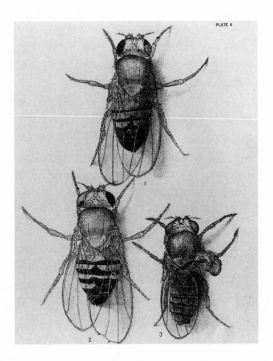

*Sexual mosaics. Notice the mismatched eye colors and wing sizes. From Sturtevant, "Origins of Gynandromorphs."*

To Benzer, gynanders were novelty-shop items. He once made a gynander with one good eye and one blind eye and put it into a vertical tube with a lightbulb overhead. A normal fly with two good eyes will climb straight up a tube toward the light. It will also climb up if the room is dark, because *Drosophila* has a sense of gravity as well as a sense of sight. But if Benzer turned on the light, his gynander would climb the tube in a corkscrewing path, because it would keep turning itself to one side, cocking its bad eye toward the light, trying to balance the input from the two sides of its head. If its right eye was bad, the gynandromorph traced a right-handed helix; if the left eye was bad, the gynandromorph traced a left-handed helix. "Sometimes," Benzer wrote, "it is difficult to resist the temptation, out of nostalgia for the old molecular-biology days, to put in two flies and let them generate a double helix."

Decades earlier, Sturtevant had realized how to use gynanders to make a form of what embryologists call a "fate map," a map of the fate of every point in the early embryo, showing which part of a fly develops from each cluster of cells. At a very early embryonic stage called the blastoderm, a fly egg is covered with a smooth layer not of shell but of cells, ten thousand cells. In a gynandromorph embryo, half of these cells are female and half male. Gynander eggs are like Easter eggs in which each and every egg has been hand-dipped into two bowls of dye. The sexual boundary line may run crosswise, slantwise, any which way across the gynander egg, but it divides the surface into two parts, male and female.

Sturtevant had once worked out a way to trace the male and female parts of each gynander back to its point of origin on the surface of the blastoderm. To try out his idea, he had examined 379 gynandromorphs of *Drosophila simulans,* a close cousin of *melanogaster.* He had drawn pictures of each one, and noted which part of each was male and which female. Then he had put these sketches away and gone on to other projects. In 1969, two fly men—one of them a postdoc of Sturtevant's student Ed Lewis—borrowed that now yellowed sheaf of drawings and used them to finish the project. They drew an oval map of the blastoderm's surface and plotted the point of origin of the first left leg, the second left leg, and the third left leg; the head, the eyes, and the wings; the sections of the dorsal and ventral abdomen.

Sturtevant was dying while they worked on the fate map. They finished it just before he died.

Next, Benzer and Hotta made a fate map of the *melanogaster* egg. Like many of Sturtevant's friends, they had always regretted that the map unit of genetics is named after Morgan and not after the man they called Sturt. To honor Sturtevant's memory, Benzer decided to measure distances on his fate map in *sturts.* The distance from the point of origin of the fly's first left leg to the point of origin of the fly's second left leg, for instance, is ten sturts. Benzer could not quite pronounce the name of the new map unit without giggling. "But, you know," he says, "that was a sentimental thing with me, naming it after Sturtevant."

Now Benzer's postdoc Jeff Hall began using gynandromorphs and fate maps to explore sexual instincts. He and another postdoc in the Benzer lab, Doug Kankel, knocked out a gynander with ether, mounted it in a kind of white goop (one brand name is Tissue-tek), and deep-froze it. Then, with a microtome, they sliced it into thin sections—so thin that they got as many as thirty or forty slices from a single fruit fly. They stained each section in such a way that the male cells stayed colorless, while the female cells turned dark brown. By looking through the microscope at these stained sections of the fly's brain, Hall and Kankel could see which pieces of the nervous system were male or female, right down to individual neurons. In this way Hall would later map the portion of the brain that has to be female if a fly is going to elicit courtship from a male,

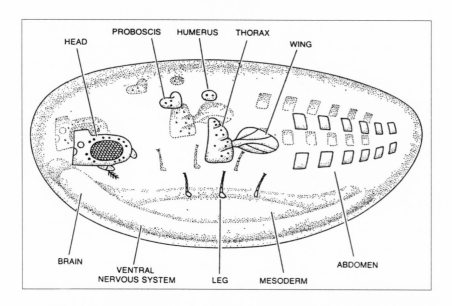

HEAD  PROBOSCIS  HUMERUS  THORAX  WING  BRAIN  VENTRAL NERVOUS SYSTEM  LEG  MESODERM  ABDOMEN

*A fate map. This egg-shaped object is a fly embryo at the early stage called the blastoderm. Benzer's diagram shows how each part of the adult fly comes from a specific site on the blastoderm's surface. After making this fate map, he and his students traced pieces of behavior back to the surface, too, including many of the dance steps of courtship and copulation. From Benzer, "Genetic Dissection of Behavior."*

and he also mapped the portion of a fly that must be male (a spot in his midsection) if he is going to try to copulate with her. Even if the cuticle of a gynander's head is female, with female eyes, female ears, and a female head capsule; even if she has a female thorax and a female thoracic ganglion; and even if her wings are female, she will still hold out her wing and make it tremble in the love song of the male fly if just one critical focus inside her brain is male. On the other hand, a gynander with a body that is almost all male will still be receptive to a male's song and dance if just one critical focus in its brain is female.

Benzer could add each of these pieces of behavior to the fate map the same way he and his students mapped each piece of anatomy. Each week, their egg-shaped fate map was decorated with more landmarks of anatomy and behavior, all inscribed in the same spidery, dead black lettering that Benzer had used since *Arrowsmith*.

The project mattered as a way into the strange territory between genes and behavior, territory that would come to absorb Benzer more and more deeply. In a sense, the territory defined what was original about his project. Animal breeders and ethologists could see that instincts pass on from generation to generation. But only molecular biologists could go inside and see how the path from gene to behavior begins inside the embryo.

At that time the way an embryo grows and develops (a subject known to biologists simply as development) was still a complete mystery. The rules and origins of development were as obscure as the rules and origins of behavior. No one knew where to search for answers, and no one even knew what the answers should look like. Genes linked with the early development of the embryo had been on the maps for sixty years by then, but the problem was essentially imponderable. No one realized that the genes that Ed Lewis was mapping in Sturtevant's old lab space would crack open the problem.

Lewis had arrived at Caltech in 1937 to write a Ph.D. thesis under Sturtevant. Ever since 1937 he had been sitting in the same Fly Room working on a few particularly bizarre lines of mutants. One was *bithorax,* which has an extra set of wings: a four-winged fly, first discovered in a half-pint milk bottle in 1915. Another mutant was *Antennapedia,* which has legs growing out of its head where its antennae should be. By the 1970s, after examining and crossing hundreds of thousands of deformed mutant flies, Lewis had begun to understand something about the roles that *bithorax* and *Antennapedia* play in the fly's development. These genes control the body plan of the back half of the fly, everything below the head. Slowly, without publishing much of his research, Lewis mapped the genes in a great mural on his lab wall, and generations of Caltech administrators let him keep at it. "In any other institution—" Benzer would declare decades later at a rollicking campus party in Lewis's honor, in front of snapping, clicking, and whirring cameras, after the world had finally caught up with Lewis's work. "In any other institution—" By then Lewis had long since retired from teaching, but Caltech had allowed him to keep his old laboratory. He still worked almost as hard as ever as the Thomas Hunt Morgan Professor of Biology, Emeritus. Only a school devoted to pure research and to the memory of T. H. Morgan would have allowed Lewis to peg away decade after decade while he mapped *bithorax.* It was a project that sounded quintessentially irrelevant and proved central—like Benzer's.

In a normal fly, the thorax has three segments. The first segment has a pair of legs. The second segment has a pair of legs and a pair of wings. The third segment has a pair of legs and a pair of balancers, called halteres. Lewis discovered that *bithorax*'s problem is not a single mutation but a cluster of mutations on the third chromosome. In a fly embryo these mutations confuse the identity of one segment with the identity of the next. When Lewis mapped them patiently decade after decade, he found that the genes in the *bithorax* complex are arranged in the same order along the chromosome as the parts of the body they affect. That is, if one goes down the chromosome from top to bottom, one comes to genes that affect the growth of the fly from the top of the head to the tip of the abdomen. The genes that control the development of the head and the antennae are at one end of the complex, and the genes that control the development of the tip of the abdomen and the anus are at the other end. What is more, these genes turn on one after the other as the fly embryo is growing, in anatomical order, beginning with the head and working down toward the anus; and if they switch on in a different order or if one of them misfires, the body plan of the fly is disarranged. In this sense, the fly itself is a map of its genes. As one drosophilist writes, "It is as if the insect's entire body is the expression of a giant chromosome made visible to the naked eye."

This complex of genes would eventually lead molecular biologists inside the problem of development the way Benzer's mutants would lead them inside the problem of behavior. What Lewis found in flies would turn out to be fundamental throughout the tree of life. New tools of molecular biology would augment the old tools of genetics and mapping to produce breakthrough after breakthrough; and the same tools would work equal wonders with Benzer's mutants. Clock mutants and the *savoir-faire* mutants would provide the first set of picture windows into the workings of genes and behavior at the level of the anatomies of atoms.

In the last years of the century, Lewis, standing by his old teacher Sturtevant's iris bed or besieged by reporters in the faculty parking lot, would say with a grin, "It was pure genetics. It was pure genetics." Nothing molecular about it. He would remember how Delbrück had pounded the table and denounced the fly, and he would murmur, so softly that the reporters had to ask him to raise his voice, "I'm glad I stayed with it."

MEANWHILE, Jeff Hall accumulated a shelf of bottles of mutants with interesting courtship problems. One mutant male courted vigorously but never copulated. Hall named him *celibate*. Another male mutant disengaged after only about ten or twelve minutes and rarely fathered any children. Hall named him *coitus interruptus*. Then there is *stuck*, which was discovered in another fly lab. A *stuck* male has trouble withdrawing his penis after copulation. "The pair just sticks together," Benzer says, "and they keep pulling against each other for hours or days on end. Sometimes they die of starvation."

In the wild, of course, a mutation like *stuck* could not last long. A male that is *stuck* will not pass on his genetic inheritance. But this particular mutation is recessive, which means that flies with one mutant copy and one normal copy of the gene can court and mate normally. In the Church Laboratory, Benzer and Hall began collecting and breeding these courtship mutants, keeping them going from generation to generation like seedless grapes.

The most surprising courtship mutant was discovered at Yale. A graduate student in a Fly Room there was studying the process of egg formation. He zapped flies with X rays and looked for sterile female mutants. In his fly bottles he happened to notice a few mutant males that were courting each other. Ordinary males will bump and bustle and dodge around other males in a fly bottle or a petri dish without ever touching one another for long. Watching them is like looking down at a sea of strangers in Penn Station. If two males do collide in the crowd, they may retreat a step or two and begin grooming themselves, almost as if they were embarrassed, like two cats. If a normal male sees another male approaching him with a wing out, he starts flicking his own wings violently in rejection. But these males at Yale sang to each other. Tests showed that their sperm was healthy and that they would court females but would not copulate. They would go through all the steps of foreplay, but the last step they would not do.

Sometimes three males, or five, ten, or more would form chains and follow each other around in the fly bottle in long, winding conga lines. They would chain for hours. They tended to stay down around the food at the bottom of a bottle, but when the dancing had reached a certain pitch of frenzy they would get right up onto the glass walls of the bottle and chain. Often they broke up and then came back together. They took little breathing spaces and went right back to chaining.

Those who have worked with this mutant have found that food is important: Giving the flies good food and keeping them at a nice warm constant temperature seems to encourage chaining. "Sometimes it takes a few days," one technician says. "It's like a *social* thing. They all get to know each other." The food, the climate, and perhaps the social environment have to be right. "When they are really happy," she says, with a half-apologetic smile for talking about the happiness of flies, "then you get these chains."

The investigator at Yale published a note about this mutant, which he named *fruity*. Then he went back to his home country, India, where he was never heard from again, and *fruity* was left an orphan. When Hall read about it, he decided that this was the quintessential courtship gene. He sent to Yale for a bottle of *fruity*, although he decided that the name had to go. He chose *fruitless*.

In Dante's vision of the tenth circle of Hell, sodomites go round and round with their bodies linked in a wheel, circling on burnt sand in a whirl of ever-moving feet. In Benzer's Fly Room, the scene looked like Dante's. The male mutant flies whirled in bottles and petri dishes and test tubes—long swirling sinuous chains, males only, winding their way around and around, hour after hour. The males never tried to copulate. They only formed these long conga lines in their milk bottles and test tubes and danced, sometimes sticking out one wing and singing the love song of their kind while they danced, as if they were shaking a tambourine.

In the 1970s, most aspects of sexual behavior were still a black box, at the level of atoms, molecules, and genes. Biologists had collected observations of sexual behavior in tens of thousands of species. The importance of the instinct, which is after all indispensable, and the phenomenally variable behavior of closely related species like Hawaiian *Drosophila* suggested that in most cases every step of the instinct must be inherited. And for the atomic theory of behavior, *fruitless* would provide the first way in.

The *fruitless* males chained on the floor of their bottle. They chained on the sides of the glass in big wheels and zeroes all day long, more and more of them spiraling up the walls toward dusk. And when Jeff Hall dumped them out of their bottle, he could see them chaining all the way down the funnel, chaining into the morgue.

# CHAPTER TEN

# First Memory

*Memory is a passion no less powerful or pervasive than love.*

—ELIE WIESEL,
*All Rivers Run to the Sea*

BENZER'S FRIEND Gunther Stent loved to philosophize, and Benzer was always quick to help him get to the bottom line. Once Stent inhaled for what was clearly going to be a long verbal essay. He began, "All reasonable men—"

Benzer cut in: "No men are reasonable."

Benzer and his circle had an interest in the truth and a cavalier impatience with mouth artistry and eloquence, and their attitude, their style of science was extremely appealing to a certain kind of student. It is an old saw in organic chemistry that "like dissolves like." That is the principle of the countercurrent method by which the chemist separates oil-lovers from water-lovers. In the first days of their genes-and-behavior projects, Benzer with his flies and Brenner with his worms attracted some students and repelled others. Those who gravitated to them were a special breed, looking for a certain kind of adventure. Like Benzer and Brenner, they hated the safe, careerist path. "There were a lot of young people who were already feeling that molecular biology and genetics had reached a kind of confined road," Brenner says. "In other words, they didn't feel there was opportunity for originality or initiative anymore. So we attracted a large number of people who were willing to take the chance and enter the field. And, of course, those are the people you want." Most of their students were coming to them from classical phage genetics. "And for a lot of those people that was quite a big step into the void."

Those who visited Benzer's Fly Room found an atmosphere not unlike Morgan's Fly Room, which had been a madhouse. Morgan had wedged eight desks into his little room at No. 613 Schermerhorn Hall for his Raiders. Escaped flies drifted around the desks and the garbage can, which was never completely empty. More flies swarmed around a bunch of bananas that hung in the corner. When the boys came into the room in the mornings, bearing stolen milk bottles, they would pick and eat bananas and drop the peels into their desk drawers. "During the two years I worked in the inspiring spiritual atmosphere of the Fly Room," wrote one of Morgan's finest, Curt Stern, "I never opened my desk drawer without looking away for a while to give the cockroaches a chance to run into the darkness. Once I said breathlessly, 'Dr. Morgan, if you put your foot down you'll kill a mouse.' He did!"

"Morgan was a bit crazy," says Jeff Hall. "He used to say, 'To know your organism, you must eat it.' Not just the flies: the pupae. And not just to horrify people, but to *know*. Grape Nuts had just been invented, by C. W. Post. One of the first breakfast cereals." (Post put Grape Nuts on the market in 1897.) "So Morgan justified eating the pupae—when people stared at him, he said, 'They taste like Grape Nuts.'"

Benzer brought to his own Fly Room a growing fascination with the bizarre, the extremities and ultimities of human behavior. "The whole lab was infected by his spirit of the unusual," says Bill Harris, who was a graduate student in those early days and is now the chairman of the department of anatomy at Cambridge University. Benzer took his students and his wife, Dotty, to the Charles Manson courtroom. He visited the grave of Marilyn Monroe and crashed Hollywood funerals. And of course Benzer often brought surprises to his lab's lunchroom, known locally as Seymour's Sandwich Shop. Harris remembers in particular a Chinese century egg, an egg that had been buried in the ground for years. The white of the egg was a translucent red, and the yolk a very dark green. "He ate it up and made everyone taste some. My old boss—always liked to challenge his palate."

For Benzer it was all part of the same curiosity: "All part of the same aberration," he says. And his students saw it that way too, according to Harris. "That was one of the most attractive things about his science. I didn't know him in his early years, when he was working on the gene. But this was definitely science at the fringe." There was so much wide-open space between a mutation and a piece of behavior. Benzer was always quoting Samuel Butler: "A hen is only an egg's way of making another egg." Behavior, Benzer said, is "the way that the genome interacts with the outside world," the way the egg produces that next egg. Benzer's approach, getting a mutant and looking at a behavior, underscored how little was known about the whole circle from the egg to the winged life and back to the egg. "There was so much uncharted territory in between," Harris says, "that most scientists believed it was an unfillable gap. And most of us were attracted to it because of that." They did not know that the tools of molecular biology would soon allow them to follow up these first explorations with fantastic power, allowing them to study the links between genes and behavior at a new level. "But we knew that it was a process, that little by little we would inch away at this problem," says Harris. Precisely because the gap was so wide and because many scientists thought the project was absurd, it attracted a certain kind of student. "It was not a step," Harris says. "It was kind of like a leap."

"The lab in those days was very laissez-faire," says Chip Quinn, who joined it as a postdoc in 1971 and now runs a Fly Room at the Massachusetts Institute of Technology. "There were these fairly interminable lunches, which were sometimes movies and bullshit and sometimes real science. I think at that time in the lab nobody knew exactly what should be done. There was this whole cafeteria of things that one might do." Quinn remembers one student who made the rounds of Benzer's three-ring fly circus looking more and more disgusted. "He said, 'Well, here I am training for Harvard Neurobiology and it's sort of like all these guys with coonskin *caps*. And you want me to come here and put on a coonskin cap and be a *pioneer*.'" That student went off and did something safer.

No one knew how important these experiments would turn out to be. "Nobody knew," says Quinn. "Seymour always thought, you know, enlightenment is just around the corner. He really had confidence in himself as he came off the big successes in phage, and he thought, 'Well, we can do it, we can figure out the nervous system.'" To Quinn it was a noble gamble, like Pascal's wager. "If you think you can't do anything, then you are guaranteed that you can't do anything," he says.

OF ALL THE PROJECTS in the lab, Quinn's was the farthest out. Quinn wanted to dissect the invisible events that take place in the brain during and after each experience, the changes we call learning and memory. Quinn even hoped to find the engram, which is the holy grail for scientists who study nerves and brains. The engram is the seat of memory, the physical change in a brain that encodes memory itself. "Tell me what is a thought, & of what substance is it made?" asks William Blake, as if he were asking an unanswerable question. The engram is the substance of memory. In 1971 it was still a cloud-wrapped, faraway summit, but Quinn thought it might turn out to be easier to reach than some of the other summits on the horizons of science. "What's the trick or the set of tricks that the brain uses to encode a change based upon experience?" he asked years later, after his own experiments and those of many others had brought them closer to the summit. "It may be relatively simple. I mean, there's really that hope: the brain is too complicated, and intelligence is too complicated, and consciousness is too complicated, but there is a possibility of understanding *this trick*."

This is the trick Quinn wanted to study using the tools of genetic dissection, the trick that allows us to catch something from our experience in a kind of mesh of the nerves and hold it there for the rest of our lives. Somehow the memories are written in atoms, and somehow we keep the memories even though we lose the atoms.

And of course there is a wide range from individual to individual in our ability to remember. The psychologist A. R. Luria wrote a famous memoir of his encounters with a newspaper reporter who was sent to him by the reporter's editor because the man seemed to forget nothing. Luria sat him down and read out a short table of numbers, and afterward the reporter repeated them back to him digit by digit.

$$\begin{array}{cccc} 1 & 6 & 8 & 4 \\ 7 & 9 & 3 & 5 \\ 4 & 2 & 3 & 7 \\ 3 & 8 & 9 & 1 \end{array}$$

The psychologist read him longer and longer tables, row after row, stream after stream of digits, and the reporter kept repeating them back. He could recite them backward, forward, or even diagonally. At last, the psychologist writes in his memoir, *The Mind of a Mnemonist*, "I simply had to admit that the capacity of his memory *had no distinct limits;* that I had been unable to perform what one would think was the simplest task a psychologist can do: measure the capacity of an individual's memory."

This subject had a special interest to the Lords of the Flies in the Benzer lab because many of them had bottomless memories themselves. They needed them for their work. Ron Konopka was born with a photographic memory. Jeff Hall carried in his head thousands of references to papers in genetics, and often he could remember not only the authors of a paper and the genealogies of the flies but also the year, volume, and page numbers. Sturtevant used to read the *Encyclopaedia Britannica* for pleasure in the evenings, and in his later years he had a hard time finding an article that he had not already read and committed to memory.

This is the phenomenon that Quinn wanted to explore through genetic dissection. He also hoped to explore the way the dance of the atoms changes over time. As we get older, most of us feel the holes of the sieve widening and widening, the mesh fraying, so that more and more of what happens from morning till night falls through. This change in our daily experience probably corresponds to a change in the sieve and mesh of the molecules and neurons of memory. Something changes in the way we store memories or in the way we retrieve and read them. Quinn hoped he could find clues to those changes too with the dissecting needle of the genes.

"For us to learn anything at all, we must already know a lot," says another early postdoc of Benzer's, Yadin Dudai, who now runs a Fly Room at the Weizmann Institute of Science in Rehovot, Israel. We have to know how to live to know how to learn. As Dudai writes in his book *The Neurobiology of Memory,* "What the frog's eye tells the frog's brain is based on a memory established during millions of generations; so is the escape of the fly from the frog's tongue." In this very broad sense, genes themselves are ancestral memories of life on earth. Near the beginning of *Remembrance of Things Past,* Proust says that memory is "a rope let down from heaven to draw me up out of the abyss of not-being." All of DNA is a twisted rope ladder let down from heaven to draw us up from the abyss of not-being. We do not lift a finger without three kinds of information: the information we are getting from our senses at that moment; the information we have gotten from our senses in the past; and the information our ancestors have acquired since life began on Earth—that is, the information that is represented by genes themselves. Evolution is learning. Species store learning in chromosomes the way individuals store learning in their brains and societies store learning in books.

In this sense our ability to learn and remember is itself a memory. It is the memory of a discovery that has been passed down from generation to generation since near the beginning of life, a discovery approximately as old as the sense of time, perhaps almost as old as the instinct to reproduce. And of all the discoveries that living things have acquired in their 3.5-billion-year tenure on Earth, the mnemonic device of memory itself is one of the most crucial. For an individual to be able to profit from its

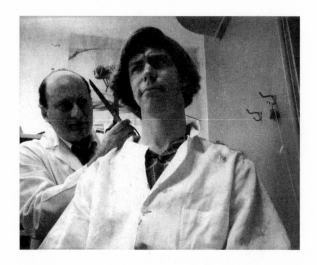

*A sentimental sense of history. Once, back in the 1940s, Benzer had traded haircuts with his mentor Max Delbrück. Now, in the 1960s and 1970s, while he explored the molecular biology of behavior, Benzer traded haircuts with his postdocs. Here Chip Quinn endures the tradition, in June 1974.*

experience and carry each experience forward to the next choice point is one of the most useful adaptations ever evolved.

Quinn had picked up an interest in the memory problem from Benzer, but he felt almost mystically primed for it. As Quinn puts it, "I think that it was my karma as well as his." To Quinn, the engram, the secret of memory, was by far the most exciting problem he could hope to study in Benzer's lab. "Everything else seemed trivial compared to that."

AT THAT TIME, flies were assumed to be hard-wired. That is, every one of the hundred thousand neurons in their brains was thought to be glued, taped, or soldered to its neighboring neurons in a pattern that was laid down once and for all in the embryo. The layout of the nerves was supposed to be as fixed and standardized as the layout of the six legs and two wings. A fly's brain never changed no matter what happened to that fly between the moment it eclosed from the egg and the moment it met its maker (or mutater). Without memory, John Locke once wrote, each of us would be no better than "a looking-glass, which constantly receives a variety of images, or ideas, but retains none; they disappear and vanish, and there remain no footsteps of them; the looking-glass is never the better for such ideas." A fly was thought to be as unimpressionable as a looking glass. All it had to fly on was the ancestral memories in its genes: a pure robot, a set of instruments flying on instruments. Students in Benzer's lab sometimes wondered if worlds exist where there is no other sort of memory but this, the slow, instinctive memory built up out of thousands and millions of generations. It was an amazing thought. What would such a world be like? Billiard balls, pure billiard balls! A planet almost as barren as a planet without life. Our own planet in the days of the very first living forms must have been like that, before an organism somewhere in the sea learned to profit even dimly from experience—learned to learn.

For years a fly man at the University of Pennsylvania, Vincent Dethier, had suspected that even flies could learn. Where most people see flies as "little machines in a deep sleep," Dethier once wrote, he looked through the microscope at their fantastically intricate armored bodies, "their staring eyes, and their mute performances," and could not help wondering if there might be "someone inside." Dethier tried to prove that flies can remember but he never could, and after eighteen years he gave up. Today when Benzer lectures about genes and memory, he often flashes on the projection screen an old headline from the *Washington Post* about Dethier's conclusions. The *Post* ran a rather unflattering close-up photograph of the face of a fly with a caption that passed the finger-shaking judgment "Can't learn anything."

In those years Benzer taught an undergraduate course in behavior at Caltech. He always put the same question at the end of the final exam. The question was worth a case of beer and five hundred points (which meant an A-plus for a satisfactory answer): "Design an experimental situation in which you can show that *Drosophila* learn." Many young techies rose to the challenge. One graduate student in Benzer's laboratory arranged a tiny spotlight so as to cast a shadow of the fly on a sensor. By administering a series of little punishments in the form of heat, the student, Jeff Ramm, tried to train the fly to alter its posture. After a while he got discouraged and left the laboratory. Benzer used to complain that Ramm had quit just before the flies learned. "But I think that Jeff and Seymour had incompatible personalities," says Chip Quinn. "Jeff Ramm was someone who didn't have this tolerance for starting experiments when you don't know what you are doing. And so he diffused off to Felix Strumwasser's lab." Strumwasser was a neurophysiologist, an expert in the workings of nerves. He had been one of the biggest naysayers when Benzer presented his genes-and-behavior project to the Sperry group.

Another student of Benzer's wrote a paper proposing to adapt a cockroach experiment called Horridge Leg Lifting. A student of invertebrate behavior, Adrian Horridge, had stuck a cockroach on a tiny diving board. If the cockroach's legs dangled into the water, it would get a shock. Eventually the cockroach learned to lift its legs out of the water. Benzer's student thought Horridge Leg Lifting might work with flies. But he quit the lab and went off into computer science. "Again," says Quinn, "I think he quit just before they learned."

Benzer himself tried to teach flies in his countercurrent machine. He put an electric grid inside one of the test tubes and shocked the flies to teach them to stop going toward the light. If he shocked the flies over and over again, they would go toward the light less and less. When he turned off the current, the flies still did not run toward the light. For one brief happy period, Benzer thought he had taught his flies to avoid the light. But when he put them into a fresh test tube, they ran for the light as urgently as ever. Apparently the flies were not learning after all. They were laying down some sort of odor—perhaps the odor of panic, or the odor of singed fly hairs and feet—and avoiding a bad smell in a test tube is not the same thing as learning; it does not imply a remembrance of things past.

When Quinn joined Benzer's laboratory, he looked over the countercurrent machines and shock grids that Benzer had made. "And at some level, I had no idea what to do," Quinn says. So he started simplemindedly repeating Benzer's experiments and replicating his results, "just to see what was going on, because I didn't know what to do." He found that the flies did seem to be stinking up the test tubes. But he also found that if he took fresh flies—naive subjects, unshocked troops—and ran them through the countercurrent machine, these new flies would ignore the odor and head for the light as if the odor were not there. Quinn concluded that the odor of panic (or whatever it was) repelled the flies only if they had smelled it in conjunction with an electric shock. In other words, Benzer's flies might have learned something after all: they might have learned to avoid the odor. "So that looked relatively encouraging," Quinn says.

Since the flies apparently paid some attention to odors, Quinn (guided by Benzer) decided to try perfuming one of Benzer's countercurrent machines. "Caltech had a whole room full of old bottles of chemicals. So I went around and opened them and sniffed them." Quinn had no way to know which odors to pick, because he did not know if the odors would smell the same to flies as they smelled to him. On the other hand, he had no better way to pick and choose than to please himself. He wanted a compound to be volatile enough that he could smell it, but not so volatile that it would go away immediately. So he went around the room opening vials at random. He chose a vial labeled "Octanol," which smelled like licorice. He also chose a vial labeled "Methylcyclohexanol," which smelled—as someone in his lab later put it—"a lot like tennis shoes in July."

Quinn lined the test tubes in the countercurrent machine with copper grids—very fine mesh, like tiny rolled-up window screens. He perfumed one test tube with octanol, another with methylcyclohexanol. The odors would linger powerfully on the grids for two hours or so. When he was ready to start the experiment, he laid this set of test tubes on his benchtop, just as Benzer had done, and lit Benzer's fifteen-watt fluorescent bulb. He put about forty flies in the first tube and let them explore it for sixty seconds. Then he tapped the whole apparatus on a rubber mat to knock the flies down to the bottom of the tube (just like Benzer), as if he were making an official pronouncement in the dark room: "This meeting will now come to order." Finally he slid the tubes so that the flies could walk, if they chose, into the tube ahead of them, which smelled of octanol.

As in Benzer's first experiments, almost all of the flies walked or ran toward the light—where they encountered a shock of seventy volts for fifteen seconds, which might kill a human being but only ruffled the fly's bristles a bit. Then Quinn tapped the tubes on the benchtop again to return the flies to the first tube and he gave them sixty seconds to recover from the shock.

Next he slid away the tube that smelled of octanol and replaced it with a tube that smelled of methylcyclohexanol. Again the flies ran for the light. This time Quinn did not give them a shock. After fifteen seconds he returned them to the first test tube.

Quinn repeated this cycle: octanol and a shock, meth and no shock. In a way, he was talking to the flies: octanol, bad; methylcyclohexanol, good. (Of course, there was nothing intrinsically good or bad about either; he had tossed a coin before starting the experiment.)

Finally, the test. Quinn slid into place a fresh test tube, one the flies had not encountered before. The tube was perfumed with octanol. He bent over the tubes by the fifteen watts of the fluorescent light and watched. More than half of the flies milled around and did not go into the tube.

After returning all of the flies to the first test tube and giving them a minute to rest, he slid away the tube of octanol and slid into place a fresh tube of methylcyclohexanol. Most of the flies went into that tube.

It was an eerie thing to watch. The flies were acting on their experience. Ever since Morgan's Fly Room, geneticists had known that fruit flies have genes and chromosomes as we do and that they inherit their bodies and behavior as we do. But here the flies were doing something that even Quinn had not really expected to see a fly do: they were learning from their own experience. They have something inside that allows them to remember what has happened to them, as we do; and they can act on what they remember, as we do. It was a sight to give a human being pause, like William Blake's visionary line: "Am not I a fly like thee / Or art not thou a man like me?"

Quinn repeated the experiment with a second crowd of flies, and they learned the lesson, too. He tried teaching a third crowd of flies the reverse lesson: methylcyclohexanol bad, octanol good. The flies learned that lesson, too. Each time the change in the flies' behavior was eerie to watch. It was as if something palpable had changed for them, as if the odors had turned into invisible doors. They acted as if one door was still wide open but the other door was almost closed and looked forbidding to most of the flies.

To make sure he was not just kidding himself, Quinn asked a friend in the lab to set up the teaching machine for him, to arrange the perfumed test tubes. This way, when Quinn ran the experiment, he could not know which test tube held which odor and he could not know which odor led to a shock. He ran the experiment blind and recorded the flies' behavior impartially: how many chose this tube and how many chose that one. Then he checked his friend's notebook (in effect, taking off the blindfold) to see if the flies were really learning their lessons. They were.

Not only were the flies learning, they were learning fast, as Benzer enjoyed pointing out chauvinistically. In one standard laboratory test of learning, an experimenter rings a bell and then blows a puff of air at a rabbit's eye to make it blink. The rabbit learns to blink before the puff of air, but that takes about eighty lessons. Quinn's flies learned in three.

Once Quinn happened to notice that his flies avoided walking on a spill of dry powder—quinine sulfate. So instead of electrifying his grids, he tried dusting them with the powder, spreading some quinine sulfate on the copper with a fine artist's brush. The flies learned to avoid the powder the same way they learned to avoid a shock. Benzer's graduate student Bill Harris built a **Y** maze of black Plexiglas with small brass knobs. At the choice point of the **Y** maze the flies could go toward either a red light or a blue light. The flies learned those lessons too: red good, blue bad; or red bad, blue good.

They seemed to be learning much the way human beings learn: by repetition. It is adaptive for living things' memories to require repetition. A three-month-old baby will not remember something that happens once, but if the lesson is reinforced on a regular schedule, she will. "A single experience that is never to repeat itself is biologically irrelevant," the physicist Schrödinger wrote. "Biological value lies only in learning the suitable reaction to a situation that offers itself again and again, in many cases periodically, and always requires the same response if the organism is to hold its ground."

Next, Quinn played with his flies to see how long they would remember their lesson. He taught a fresh group of conscripts to avoid octanol and let them rest for an hour. Then he put them through their paces again. One hour in the lives of fruit flies is like a few months in ours. After the break, most of the flies still remembered to avoid octanol, but some of them forgot. He gave another group of flies twenty-four hours (six years in human lives) to sit around in fly bottles and forget. Many of them still remembered; more of them forgot.

"CHIP QUINN once described the ideal organism as one that has three large neurons, divides rapidly, and can learn to play the piano," Benzer says. "Everybody wants a simple system." Delbrück, Benzer, and Brenner had thought of the fungus, the fly, and the nematode worm as simple systems when they started: as something like gadgets in physics, atoms of behavior. "Sydney's idea was, the nematode has a small number of neurons; therefore, it's a simple system," Benzer says. "I think he has changed his mind a little bit. He describes the lineage and the development of the nervous system of the nematode as 'baroque.'" (This discovery was foreseeable even in the first days of the Enlightenment. "A worm is only a worm," said Diderot. "But that only means that the marvelous complexity of his organization is hidden from us by his extreme smallness.") Now Benzer and his students were beginning to realize that the fly is baroque too; they were delighted to discover that it can learn and act on what it learns.

Because the fly has memory in its repertoire, Benzer and his students could begin to dissect this behavior too. They could use their tools of genetic dissection to take apart the act of remembering the same way they were taking apart the sense of time and the dance of love. A new student at the Benzer lab, Duncan Byers, dosed flies with the laboratory's favorite mutagen, EMS, producing five hundred separate lines of mutants, and he began to test their memories, strain by strain, mutant by mutant, using the whole Quinnian rainbow of odors, including octanol, methylcyclohexanol, and quinine sulfate. Out of the five hundred lines of mutant flies, about twenty lines failed to learn. But almost all of those flies flunked other tests too. Some turned out to have problems seeing; others had problems smelling; some were sluggish or shaky walkers. Only one of the lines that flunked had normal instincts, normal senses, a healthy level of energy, but no talent for learning. That line of mutants turned up in Bottle 38. For them to appear so early in the search was auspicious. They fit Konopka's Law: "If you don't find it in the first two hundred, quit." Byers examined them through the microscope. As eggs, larvae, and pupae, they looked normal, and they lived as long as normal flies. They ran to the light, they climbed walls, they flew, walked, courted, and copulated like normal flies. But for them every odor was an open door. They never seemed to learn.

Benzer and his crew decided to name this new mutant after John Duns Scotus, the thirteenth-century philosopher. Duns Scotus's disciples were known as Scotists, dunses, or dunces. In the sixteenth century, the dunces lampooned the new learning, and they were lampooned back by the natural philosophers who were the world's first true scientists. The dunces lost the war, and the scientists made them eternal symbols of stupidity.

Benzer's raiders tried very hard to get *dunce* to learn something. They exposed the new mutants to a wide range of odors in all kinds of dilutions and combinations, and to a wide range of shocks, from 20 volts to 140 volts. But the flies kept their dunce caps. Benzer was delighted. With mutants like these, he and his students could now dissect the act of memory into a series of steps. They would search for mutants that could remember a little better than *dunce*—some for a few minutes, some for a few hours, some for a few days. If they could find them and figure out what made each one different from the next, they might be able to use *dunce* and the rest to trace the invisible steps through which an experience becomes a short-term memory and then a long-term memory. Benzer's crew mapped *dunce*. The gene lies on the far left tip of the X chromosome, just a few map units away from *white* and from the mutants of Konopka, the mutants that lost the sense of time.

. . .

BENZER'S PROJECT, the genetic dissection of behavior, was off to a strong start. He and his students had used genes to open three locked doors that had fascinated poets and philosophers from the beginning of Western thought.

"The body is but a watch," said Julien Offray de La Mettrie at the start of the Enlightenment, coining a slogan for a worldview that has lasted from his time to ours. The clockwork has been called the central metaphor of modern Western civilization. Certainly from the very beginning of modern science, we have seen the space around us as a clockwork of stars and the space inside us as a clockwork of organs, a clockwork of atoms. Benzer and his students, with one of their first stabs in the dark, had found a clockwork gene. They did not know yet if they had the key to the clock; but they hoped they had a piece of it, a point of entry into a type of behavior that is as potent for us as a symbol of science as it is basic to life itself.

"The selfish gene" is another slogan for a worldview: a catchphrase for biologists—particularly biologists who study genes and behavior—since the 1970s. A body is a gene's way of making more genes or an egg's way of making more eggs, as in the quip that Benzer loves to quote from Butler. Mutants like *fruitless* were points of entry into the universal instinct that allows genes to move from one generation to the next and go wheeling across the longest geological reaches of time, from the beginning of life to the present, a span of four billion years.

"A man is but what he knoweth," wrote Sir Francis Bacon in a third slogan, a slogan that defines for us one of the traits we most value in being human. If we could not remember where we have been, we would not profit by our experience and we would not know who we are. The mutant *dunce* was a point of entry into the mechanisms that allow each of us to accumulate histories and apply the lessons we have learned at our choice points; mechanisms whose elaboration in our species has helped to set us apart from all the rest.

Darwin's process has worked powerfully and unceasingly to shape all three of these classes of genes. If animals and plants did not have clock genes, they could not keep time with the world. They would drift in and out of day and night, living less efficiently than their competitors and sometimes falling into fatal encounters. If we did not have instincts for recognizing and winning the attentions of the opposite sex, we could not pass on our genes; we would die without issue. And if we did not have memories, we could not pass these other genes safely onward and most of us could not last a day without a great deal of help from our friends.

Time, love, and memory are three bases of experience, three corner-stones of the pyramid of behavior. Benzer and his group had found a way inside all three in the first years of their Fly Room. Like Konopka's discovery of a time-wounded fly in his two hundredth bottle, it was all surprisingly fast work. Konopka's Law is broader than it sounds. Like so much that came out of the fly bottle, the message is universal. The *Iliad* and the *Odyssey* are the greatest epics ever written. Gutenberg's Bible is the most beautiful book ever printed. Some of the world's most memorable photographs were made during the first tests of the first photographic chemicals by Joseph Niepce and Louis Daguerre. Benzer's sister's husband, Harry Lapow (who gave Benzer his bar mitzvah microscope), spent years at Coney Island snapping pictures with a second-hand Ciroflex. The first day he brought the camera to the beach, he took a photograph that was included by Edward Steichen in the "Family of Man" exhibition at the Museum of Modern Art.

So Benzer's first years in Church Hall, from the first run of his first countercurrent machine, were one long confirmation of Konopka's Law. Benzer had hardly started his study of genes and behavior before he and his raiders found the time, love, and memory mutants, mutants that seem even more remarkable now than they did then, knowing what came afterward. If Benzer had tried to hoard the mutants, as many laboratory heads do, they might have died on the shelf. But Benzer chose instead to let his students and postdocs leave with them and build careers on them. As a result, each gene opened extraordinary views. By the end of the century they were helping to change our view of all behavior, including the behavior of the human family.

# PART THREE

# Pickett's Charge

*and thus beneath the web of mind I saw*
*under the west and east of web I saw*
*. . . the coiling down the coiling in the coiling*

—CONRAD AIKEN,
*"Time in the Rock"*

# CHAPTER ELEVEN

# The Drosophila Arms

*The flies, poor things, were a mine of observations.*

—PRIMO LEVI,
*"The Invisible World"*

JEFF HALL left Benzer's Fly Room in December 1973, and he opened a Fly Room of his own at Brandeis University, just outside Boston. There he hung a sign on the door: "The Drosophila Arms." On the sign he pasted photomicrographs of the sex combs on the forelegs of the male fruit fly, which help him grasp the female. By the sign of the Drosophila Arms, Hall planned to study the *savoir-faire* mutants, the ones that are unlucky in love.

Hall's breakthrough began serendipitously when he and one of his own first postdocs, Charalambos Panyiotis Kyriacou, decided to study the love songs of mutant flies. They planned to put pairs of flies inside a tiny recording studio, tape their love songs, and analyze what is known to students of fly behavior as the song's interpulse interval, or ipi.

In the scientific literature on *Drosophila* behavior, *D. melanogaster* was reported to sing with an ipi of thirty-four pulses per second. *D. simulans,* a sibling species, was reported to sing with an ipi of about twenty pulses per second. To a human ear both these songs sound alike. But if a female *melanogaster* hears thirty pulses per second she says, "Aha, that's a male of my species." If she hears the slower version, she says, "Ahh, you're a male of the wrong species, go away."

Before Hall and Kyriacou began testing mutant fly songs, they decided to measure the normal ipi for themselves. So Kyriacou built a recording studio, two centimeters long, one centimeter wide, and a third of a centimeter high, with a microphone two millimeters beneath the floor. Then he taped a series of normal male fruit fly love songs and played each tape into a machine that transformed it into a long, scrolling paper graph of pen squiggles.

Kyriacou might have recorded each song for a few seconds and taken his measurements from that. But instead, to be thorough, he and Hall recorded a full five minutes of tape, which generated a staggering amount of nervous-looking paper. Kyriacou would lug these scrolls home in the evenings, unroll them across his living room floor, and measure the ipis of fly songs while American sports events roiled across his television screen. Kyriacou is Greek by ancestry and British by citizenship, but while living in Boston he had pledged some allegiances to the local teams.

He soon found that the ipis in his scrolls agreed neither with the scientific literature nor with one another. The ipis kept changing from one end of a scroll to the other. That is, the beat varied from one minute to the next. First the song seemed to be allegro, then largo, then allegro again. Eventually, staring at one of these erratic love songs, Kyriacou decided to quit trying to measure the intervals between pulses and instead measure the intervals between the tempo changes. He was astonished to discover that this interval was regular: the tempo of the fly's song changed once every minute. "And I looked at another one, the same thing," Kyriacou says now, "and another one, the same thing." The tempo changes were not erratic: They were a secret rhythm in the song, a hidden pattern marked in time by every singing fly.

Over lunch the next day, Kyriacou told Hall what he had found. Together they wondered what the songs of Konopka's time mutants must be like. What kind of tempo changes would a fly make if it had a warped sense of time? Hall wrote to Konopka and asked him to send him some of his mutants. Konopka had published very little about *period* since he and Benzer had announced their discovery in 1971, and Hall wondered briefly if he would let them study *period* at all. "These mutants in principle were *pure gold*," Hall says. "So if he'd wanted he could easily have just withheld them and not permitted anybody else to work on them." Fortunately, fly people still shared mutants freely, a tradition that Morgan had started in the first Fly Room (the tradition has eroded recently in the gold rush that Hall's own work helped to start).

Konopka mailed Hall a few test tubes full of his clock mutants. (Fruit flies can survive for days in a test tube with food at the bottom and a cotton stopper at the top, cushioned in a well-padded envelope.) One by one, Kyriacou put the mutants into his recording studio, and in the evenings he unrolled their love songs one by one onto his living room floor. From the very first scrolls, he saw that these mutants' warped sense of time also warps their songs. Konopka's short mutant, the nineteen-hour one, for example, changes tempo much more quickly than normal. Konopka's long mutant, the twenty-nine-hour one, changes tempo much more slowly than normal. And Konopka's insomniac fly, the fly that has no sense of rhythm at all, changes tempo just as randomly as he sleeps and wakes.

These differences were obvious in the scrolls. Hall and Kyriacou also found by experiment that *simulans* females prefer a male that sings with a fast rhythm, the rhythm of their own species, whereas *melanogaster* females prefer the sixty-second rhythm. The female fly is listening closely.

Rhythm also commands attention in our own speech, of course, although we are usually no more conscious of it than is any other species on the planet. In *Lincoln at Gettysburg,* a book in Hall's extensive Civil War library, the critic and historian Gary Wills notes that throughout Abraham Lincoln's famous speech there is a strong repeating rhythm: "Triple phrases sound as to a drumbeat." Wills prints these triple phrases in a kind of accidental poem that seems to echo through our memories:

> *we are engaged . . .*
> *We are met . . .*
> *we have come . . .*
>
> *we can not dedicate . . .*
> *we can not consecrate . . .*
> *we can not hallow . . .*
>
> *that from these honored dead . . .*
> *that we here highly resolve . . .*
> *that this nation, under God . . .*
>
> *government of the people,*
> *by the people,*
> *for the people . . .*

Orators and storytellers have this gift. Somehow they communicate in the rhythm of a speech or story a human urgency, a message: hear, attend, gather round. They have the gift of producing the rhythm, and most of us have the gift of hearing it. Apparently flies have these gifts too.

Since *melanogaster* and *simulans* sing with different rhythms, Kyriacou wondered what would happen if he crossed these two species of flies. So he bred them and put their young flies in his recording studio. He found that he got two kinds of hybrid males. If his mother was a *melanogaster*, a male sang with a *melanogaster* rhythm. But if his mother was a *simulans*, he sang with a *simulans* rhythm. Since each male fly got his X chromosome from his mother, the difference in the genes was somewhere on the X.

Hall, of course, had been in Benzer's laboratory when Konopka mapped the *period* gene, the first piece of complex behavior ever mapped. So Hall knew what Kyriacou's cross might mean. The two men now knew that the gene that affects the tempo changes in the fly's love songs is on the X; and they already knew that the *period* gene is on the X. So they had to wonder if both pieces of behavior might be shaped by one and the same gene. Hall still gets excited when he talks about the moment when Kyriacou's cross pointed to the X.

"Now, where is the *period* gene of Konopka?" Hall shouts. "It's on the X chromosome! *Tah-dahm!*"


HALL EXPLAINED their hunch to his best friend at Brandeis, a young molecular biologist named Michael Rosbash. In those days, Hall, Rosbash, and Kyriacou spent a lot of time together. "And not only because of science," Hall says now, "but because of interest in Boston sports teams—all three of us being fanatic betrayees of our Boston sports teams, particularly the Red Sox."

Hall and Rosbash also played basketball together in the campus gym. The regular players were faculty and a bunch of guys from the repair department at the phone company. They played together for years—only the grad students would come and go. Sitting in the locker room day after day, Hall would talk about behavior, Konopka's gene *period*, and the secrets of the sense of time. Rosbash would talk about molecular biology.

By now the molecular revolution had begun the explosion that still continues today. A new generation of molecular biologists was doing work that was in the best sense derivative—derived from what had come before. They were adding story upon story to Occam's Castle. They had now figured out how to turn Delbrück's starting point, a petri dish of *E. coli*, into such a sophisticated all-purpose laboratory tool that even some of the young molecular biologists themselves were alarmed at how much they could do.

When a single bacterium divides in a dish, one becomes two, two become four. Within a day, by the implacable logic of exponential curves, the first bacterium has become several billion. Each cell is an identical twin of the one before. Each is what biologists call by the Greek word for a twig or a shoot: a clone.

When a particle of phage attacks a field of these clones, it injects its own strand of DNA into its victim like a hypodermic needle. (Phage particles actually look like hypodermic needles. The first time a phage watcher saw them in an electron photomicrograph, he slapped his forehead: *"Mein Gott!* They've got tails!") Sometimes a bacterium is able to attack the incoming viral DNA with special enzymes, like sabers slashing tentacles. But if these enzymes fail, the viral DNA inserts itself into the bacterium's ring of DNA, which is a circlet about a millimeter long. Once the cell receives that viral DNA, its behavior is transformed and it becomes a cloning machine at the service of the virus.

Bacterial cells also trade DNA peacefully among themselves. A bacterium will take in a smaller circlet of DNA, known as a plasmid, from a neighboring bacterium, cut out a few genes from the plasmid, and patch them into its own DNA. The secret formula for resistance to a drug such as penicillin may be carried on a single plasmid. If the cells are under attack by penicillin, a cell that carries that particular plasmid will survive and reproduce. Soon the whole petri dish has copies of the formula, some because they are clones of that surviving cell, some because they have received the formula on a plasmid and patched it into their own chromosome.

In the early 1970s, molecular biologists put this bacterial behavior to work. They figured out how to harvest a bacterium's enzyme scissors or sabers and use the enzymes to cut any pure extract of DNA into ribbons of specified lengths. These bacterial tools are known as "restriction" enzymes. Whenever they encounter DNA they restrict it, or cut it, at certain specific points. Restriction enzymes are specialized or, to use one of the buzzwords of molecular biology, specific. A streptomyces bacterium, for instance, carries an enzyme known as *Sac*I. *Sac*I will cut DNA only if it finds the sequence GAGCTC, and it will make the cut only in one place, between the T and the last C.

Collecting a war chest of these enzymes, molecular biologists in the 1970s learned to cut DNA into snippets almost as deftly as the bacteria do. They also figured out how to copy phage behavior and inject snippets of DNA into a bacterium's DNA. When their victim reproduced, it reproduced that extra bit of DNA along with its own. Then that cell's

Benzer and his first disciples enjoyed working in what one of them calls "a carnival atmosphere." These are some of the slides they showed at lectures. Benzer, going cross-eyed over flies. Benzer in front of a movie poster, doing his Jeff Goldblum imitation. Jeff Hall and Michael Rosbash at a party at Brandeis University, wearing formal attire.

children and its children's children reproduced the gene. In the space of a few hours, a bacterium had become a colony of billions, containing billions of copies of the gene—cloned.

Cloning genes turned molecular biologists into genetic engineers, a term that back in Morgan's Fly Room would have sounded like pulp science fiction. Now they could excise a gene from one species, insert it into another species, and watch what it did. They could inject a human gene into a fly. First they cloned the human gene. Then they snipped it with a specialized restriction enzyme that left the DNA fragment with what are known as "sticky ends." In a vial, they mixed a solution of these sticky DNA fragments (billions of clones of the sticky gene) with a second kind of DNA fragments called "P elements," which would act as shuttles to carry the DNA of interest into the bacterial DNA. P elements are genes that do not stay in one place in chromosomes; they hop off a chromosome and reinsert themselves elsewhere. The existence of these jumping genes was first surmised by the geneticist Barbara McClintock, working with purple, white, and spotted kernels of maize, or Indian corn, at Cold Spring Harbor. McClintock spent several decades working in isolation. Most of her colleagues found her stories about jumping genes hard to understand and hard to believe. (Jim Watson used to trample through her cornfield, chasing softballs.) Then molecular biologists discovered jumping genes in *E. coli, Drosophila,* and human beings. Each one of us carries *mariner,* a jumping gene that was first discovered in the fruit fly. This gene *mariner* may have been injected by a virus into an egg or a spermatozoan of one of our distant ancestors before the evolution of the human species itself. The gene still passes from one generation to the next. We share whole families of these jumping genes with flies, including *mariner, gypsy,* and *hobo.* McClintock won a Nobel Prize in 1983, at the age of eighty-one.

So drosophilists clone a human gene, give it sticky ends, and allow it to stick to a P element. They also add to the ribbon one of the classic genes from Morgan's Fly Room, *white.* They always use the normal form of *white*—the form that confers red eyes. Then, with a hypodermic needle, a microsyringe, they inject this submicroscopic ribbon of DNA into the rear end of the early embryo of a white-eyed fly. The whole ribbon—including the human gene, the P element, and *white*—goes floating off invisibly through the embryo. For the experiment to succeed, the P element must insert itself into one of the embryo's chromosomes—not just any chromosome, but a chromosome inside one of the cells that will eventually become a germ cell, which is a cell that makes eggs or sperm.

If the P element fails to insert itself there, then the fly's children will have white eyes. But if the P element does insert itself in the right chromosome, then the fly's children will pop out of their eggs with red eyes, and they will carry the human gene. Delbrück celebrated some of these innovations in "A Valentine for NIH":

*We now use chemistry to shuffle genes,*
*Use plasmids to move man's to beans,*
*Or rat's to microbes, flies' to fleas,*
*Or yeast's to* coli, *bee's to peas.*

*All this is based on Watson-Crick's*
*Phantastic Double Helix, plus some tricks*
*That others added to this play*
*And add still more from day to day.*

Brenner and Benzer observed these new developments with excitement and some regret. The new tools were thrilling but made working with genes almost too easy. "We had grown up in a tradition, both he and I, where it was important to try and use your *brain* in doing this work," Brenner says. As pioneers, they had been forced to look into the machinery of life from a certain distance. They had devised Olympian experiments and then made inferences and deductions, climbing through cloud banks in their minds like theoretical physicists. "And, of course, that's the great fun of it," says Brenner, "when you can cross the bridge between the machinery and your observations by *thought.*" Today they feel that young biologists are spoiled. "The present generation—you know, to them, getting hold of a gene is what you do," Brenner says. "You get hold of the gene." You clone it.

HALL HAD NEVER learned cloning. His background was classical genetics. So Hall had the mutants and he had the scientific problem. His friend Michael Rosbash had the tools, and Rosbash thought the fly might be ready to go molecular. Rosbash also thought that if he were going to move in on the fly, he would want to focus on a gene somewhere near *white,* because geneticists had been mapping and remapping that terrain ever since Morgan's Raiders. Now *period,* of course, is right next to *white.* If Hall was right about his gene, Rosbash would be able to use the tools of molecular biology to dissect an instinct: to explore for the first time the molecular links between genes and behavior.

In the locker room, after their basketball games, Hall predicted (at top volume) that *period* would turn out to be one of the most glamorous genes ever discovered. It affects everything about a fly's sense of time, from the hour of its rising up to the hour of its lying down, including the intimate rhythms in the vibrato of a love song. It is the quintessential behavior gene. But Rosbash, who is as forceful and intense as Hall, was not convinced. Rosbash thought that in spite of Benzer's, Konopka's, Hall's, and Kyriacou's enthusiasm, it was still quite possible that *period* would turn out to have nothing to do with the clock, or with the love song either. "The cell has to get up and brush its teeth, have its O.J., and so forth," Rosbash used to tell Hall in the locker room. Each cell requires vast quantities of mundane molecular machinery just to keep itself running. Rosbash thought the *period* gene might turn out to run some boring housekeeping chore without which the cell could not function smoothly. If so, a defect in the *period* gene might affect every aspect of the fly's behavior without being central to the clock. A man with the flu may not rise and shine at his normal hour, and he may not sing in the shower with his normal vigor, but that does not mean that his clock is broken. It just means he has the flu. Rosbash was raising the same objection that Benzer and his students had heard from the beginning of their project. When they poisoned flies and found lines that acted strange, how did they know they were not just breeding sick flies?

To clone *period*, Rosbash would have to invest a large amount of time in it. Although in principle the work is straightforward, the process of cloning a gene, finding out what protein it makes, and figuring out what the protein does can take years, chewing up generations of grad students and postdocs. Rosbash did not want to clone a trivial cog in some lost dark corner of cellular metabolism. The game was worth the candle only if *period* was central to the clock. Rosbash put the odds at fifty-fifty— which drove Hall up the wall.

Today when he speaks at meetings, Rosbash likes to tell the story of those afternoons in the locker room. "There was a lot of repetition," he says. He and Hall argued the same points over and over again, and so did the phone guys, who had lockers in the alley behind them. "One Monday they were telling each other about their courtship adventures in the most unfiltered terms. I'm not a prude or a delicate fellow, but their stories were extreme even for my ears. I was embarrassed for the poor girls.

"So there we were talking about fruit fly courtship, and there they were talking about mammalian courtship. And one of the phone guys came around, naked, dripping sweat, and said, 'You the molecular biologist?'"

"Yeah."

"Well, why don't we find out what the gene does?" the phone guy said impatiently. "Find out what the gene does, so we can finally get to the bottom of it and don't have to listen to this superficial description."

# CHAPTER TWELVE

---

# Cloning an Instinct

*No doubt the process of decipherment was difficult, but only by accomplishing it could one arrive at whatever truth there was to read.*

—MARCEL PROUST,
*Time Regained*

THE MOST DIRECT WAY to clone a fly gene is to take a needle and cut it right out of one of the giant chromosomes in the fly's salivary glands. These are chromosomes that carry extra copies of each gene: hundreds of copies side by side, so many that through the microscope each salivary chromosome looks swollen and banded, like an obese coral snake. When drosophilists first discovered those banded chromosomes, it was as if they were seeing their maps made flesh. In a fuzzy way they could make out the locations of *white, yellow,* and other genes that Sturtevant had mapped on that first night, and the bands were close to the positions where Sturtevant had mapped them.

Under the microscope, with a tiny needle and a micromanipulator to guide the needle, a skilled drosophilist can simply slice out the DNA from the region of the giant chromosome where the gene lies. One of the pioneers of this kind of cloning was Vincent Pirrotta, who was then at the European Molecular Biology Laboratory in Heidelberg, Germany. In 1983, at Hall's request, Pirrotta sent Hall three overlapping DNA fragments that he had cut out of the X chromosome of an ordinary, wild-type fly from positions very close to *white.*

Somewhere among these three overlapping fragments was the *period* gene. Konopka's old maps were not precise enough to say where. The gene might be on any one of the three ribbons, or it might straddle any

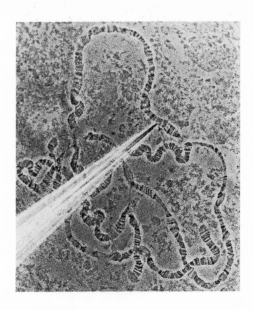

*Cloning in action. A fly has 15,000 genes. Here a microscopic needle cuts out a single one of them: the* period *gene, from the fly's first chromosome.*

two of them. To find the gene, Rosbash and Hall once again decided on the simplest and most direct strategy. They instructed students in their laboratories to make copies of the three DNA ribbons and cut the copies into random assortments of smaller fragments, using a whole battery of restriction enzymes as scissors. In this way they made a library of snipped DNA ribbons, so that they could inject each fragment of DNA into a single fly egg and test the fragments one by one.

When the DNA library was complete, they collected eggs from a mutant female fly that had no sense of time. She carried Konopka's mutant gene, the allele that makes a fly time-blind. She had two copies of that allele, one on each X chromosome. The father was also time-blind. So their children should have been time-blind too. If Hall and Rosbash could inject *period* into this family of time-blind flies and give them rhythm, they would be the first to cure a broken piece of behavior by injecting a gene. They would have inserted a gene into an embryo and given its living descendents a complex program of behavior.

Rosbash's team mixed what is known as a "DNA transformation cocktail." The cocktail included one of the mystery fragments of X chromosome from their DNA library and a P element to make the fragment restless, so that it might jump into the embryo's DNA. If the fragment of

DNA inserted itself in the right place, and if that fragment included the *period* gene, then the children of the young fly would have a normal sense of time. Each one of those newborn fly children would pop out of its egg at the normal time, the crack of dawn. Each one would go to sleep that first evening of its life at the normal time, sundown. And if the fly was a male, he would sing in standard time when he met his first female fly.

DNA injection is a straightforward laboratory procedure: wash the young embryo with ordinary household bleach (most genetic engineers prefer Clorox) to remove its chorion, the tough outer shell, then insert the microsyringe in the embryo's rear end. Of course, the embryo is so small that the injection must be performed under a microscope. One by one, the would-be engineers injected their DNA fragments into the rear ends of mutant fly embryos. According to Konopka's map, some of these DNA fragments should hold the *period* gene.

Fragment after fragment of DNA went into embryo after embryo, and students in the Drosophila Arms watched bottle after bottle. The flies they had chosen for this experiment were not only time-blind; they also had no tolerance for alcohol, because they carried a damaged form of a gene that helps flies hold their liquor. However, in the DNA transformation cocktail, Hall and Rosbash had included a normal form of that particular gene, so they could tell quickly and easily which young flies might have a clock worth testing. Any fly that had a normal ability to hold its liquor might also have a normal sense of time. When each young fly was four days old, a postdoc of Hall's, Will Zehring, put it in a little glass tube with a Kleenex soaked in alcohol. If the fly had no tolerance for alcohol, it would sip from the Kleenex and die. But if the fly had inherited the ribbon of genes on the DNA transformation cocktail, it would live. "You come back overnight and say, 'Oh, my God, we've got transformants!' " says Hall. They found about one fly that could hold its liquor for every five hundred flies that could not.

Zehring raised each individual hard-drinking fly in its own bottle, and tested it in the same way that Konopka had tested his mutants. He put it in a little glass tube, so that each time the fly walked around in the tube it would break a beam of infrared light and cause a marking pen to draw a squiggle on a scroll of paper. Then he could look at the squiggles and see when the flies were awake and when they were asleep. Hall and Rosbash kept the transformed flies in the subbasement, away from lights, in what Hall called the "pit."

In those days Hall and Rosbash were ideal partners. Hall had the genetics, the fly lore, and the flies. He also had a scrupulous sense of fair play that he had inherited from the *Drosophila* tradition. Rosbash had the drive and impatience of molecular biology and also a certain unscrupulousness that he had inherited from the molecular biology tradition. "Michael Rosbash was always a little bit like the bad boy of biology," says a molecular biologist who has known Rosbash for years. "Arrogant, always irreverent, extremely ambitious, and he earned a reputation for being a little cutthroaty—but also for being very smart." At the time he joined forces with Hall, Rosbash was working on mutants in yeast, and he had published some well-regarded papers, his colleague says. "But I always felt, when he started on the *per* gene, Michael had finally found the interesting biological story that he had been looking for." For an ambitious young molecular biologist, yeast was too crowded. "Some fields become so deep that it is hard to rise above the surf. Whereas *per* was unique. It was the only clock gene that anyone had a clue about. It was a golden problem."

Even before Rosbash and Hall began working on cloning the *period* gene, another young molecular biologist had also decided that *period* might be something wonderful. Like Rosbash, Michael Young of Rockefeller University in New York City had begun to suspect that *Drosophila* was ready to go molecular, and he also thought *period* might be the perfect place to start. In a Fly Room at Rockefeller, Young began racing Rosbash and Hall to clone *period*. He, too, collected a library of candidate ribbons of DNA and began injecting them into fly eggs one by one and watching those flies. He, too, had a big box with a roll of paper and a pen that jiggled whenever a fly moved. When he came into the lab each morning, he would roll out fifteen feet of paper and look for rhythms— the periodic bursts of squiggles in the scrolls. Working on behavior was not like doing an experiment in a petri dish, in which one sees the results in an hour or two. After each egg was injected, there would be a wait of several days. Young knew that he was racing the Boston group, and he knew that Rosbash was a speed demon. He remembers vividly how each day he would run into the lab and unroll those scrolls. In the old days, fly people had left problems alone if other people were working on them, but ever since Watson's *The Double Helix*, the ethos of molecular biology had been The Race. Young loved it. "It was movielike," he says.

Late in 1984, Hall and his students looked at their scrolls and decided they had done it: they had injected the first piece of behavior into the genes. The flies had rhythm—because they had received the instinct for it by injection. The flies would not have had rhythm unless Hall and Rosbash had succeeded in cloning the gene they were trying to clone. Hall drove to upstate New York, where the discoverer of *period*, Ronald J. Konopka himself, was now working in quasi-exile at a small college called Clarkson. Konopka had failed to get tenure in the biology department at Caltech. Hall blamed Benzer for his lack of generalship; it was one of their first serious fights. But Benzer says his colleagues were disappointed by Konopka's reluctance to publish. Konopka was a perfectionist, and he did not feel he had anything perfect to say about *period*.

At Clarkson, Konopka had built a computer setup to monitor *period* mutants in their test tubes. Before, using scrolls of paper, he had been able to monitor the behavior of half a dozen flies a week. But with his new setup, Konopka could monitor hundreds of flies each week. Hall handed his transformed flies to Konopka without telling him what it was he thought he had found, and he asked Konopka for an independent opinion of their sense of time.

"So we went home," Hall says now, "and in two or three weeks, Ron called us. He was just doing these tests blind. He didn't know what they were. He said, 'They're rhythmic.' And that nailed it. That nailed it for us. So then we knew that we had the gene." They were too full of The Race to think very much about the implications. Hall also sent some of their new mutants to Kyriacou, who by this time had established his own Fly Room at the University of Leicester in England. Kyriacou, also working blind, told Hall that the flies' love songs had rhythm. So the transformants' *period* gene was working properly in every respect: Their sense of time was solid, day by day and minute by minute.

Meanwhile—at virtually the same minute—Young and his group were testing the behavior of their own set of flies. Young's transformed flies, or transformants, had a solid sense of time too. The rival laboratories raced their papers into print. Hall and Rosbash got their paper accepted for the issue of *Cell* of December 1984. (For their help, Konopka and Kyriacou were listed on the paper as two of the coauthors.) At the same time, Young and his group got their paper accepted for the year-end issue of *Nature*—the issue that straddled December 1984 and January 1985. Afterward, Kyriacou liked to twit Young: "Well, actually, Michael, yours was '85, really."

Among those who knew what they meant, these papers caused a stir, like Konopka's announcement of the clock gene a dozen years before. In principle, what could be done with a fly could be done with a mouse or a human being. The technology of the P elements, the needles, and the markers would be much the same. Hall and Rosbash themselves have since performed an experiment that tested one of their field's futuristic possibilities, an idea that had been discussed for years both inside and outside their field with both interest and dread: Could genetic engineers someday learn to take a piece of behavior from one species and give it to another? Could they take behavior from one breed of cow and give it to another, or take some part of the temperament of one thoroughbred and give it to another, or eventually take a piece of behavior from one human being and inject it into the egg of another? Or inject a human instinct, a piece of human behavior, into a mouse or a chimpanzee?

Hall and Rosbash cloned the *period* gene of a *D. simulans*. Then they mixed one of their DNA transformation cocktails and injected it into the egg of a *melanogaster*. The piece of behavior jumped from one species to another: *mel* sang the song of *sim*.

The experiment left Hall elated: "You change the clock, and you make a different organism! We've done minievolution! We've turned one species into another!" They had transformed one of an animal's quintessential pieces of behavior. The clock gene allows it to stay in sync with sunrise and sunset—to keep time with its world, which is a piece of behavior that is essential for the animal's survival. The clock gene also allows the male fly to keep time when he courts and sings, which is essential if he is to pass on his genes. The injection had changed all that: "Not the same species anymore!"


IN THE HIPPOCRATIC SCHEME of the four humors, Hall would classify himself as choleric and melancholic. He keeps a daguerreotype of General William Tecumseh "War Is Hell" Sherman in his sanctum. (In the daguerreotype Sherman is the twin of Jeff Hall.) He also keeps a portrait of John Brown propped on top of his computer and a blue Union cap on the desk between the computer and his microscope. One of his collection of antique rifles hangs on the wall. The legend he has written across the wall in large block letters says, BE AFRAID, BE VERY AFRAID. In this context it looks like a quotation from the War Between the States. It is in fact a line from the Hollywood poster for the Jeff Goldblum remake of *The Fly*.

Hall also keeps three small dogs, terriers, in the cave under his desk. (In Benzer's Fly Room he kept dachshunds.) All day while he works, the dogs growl or wrestle with a knot of rope. Whenever a new face appears at the Dutch double doors of his sanctum, everyone in Hall's Fly Room can hear the uproar, and they can hear Hall's voice rising over it: "Now, Zoot, down! Down, down, down!"

Not long ago a reporter asked Hall a sensitive question. It is a question that Hall hears often now that the clock gene has become famous as a sort of flagship in the study of genes and behavior. Why does Hall consider clock genes a genes-and-behavior story? "Some of my friends are wondering why you call the sense of time a form of behavior," the reporter said.

Hall began calmly. "Rest versus activity is a quintessentially definable property of an organism's behavior," he said. "I think it is fair to say that sleeping and waking as instincts are not very interesting. But this is very much behavior." A fruit fly sleeps during the night, wakes up, has its breakfast, and cruises all morning, he explained. Then it takes a siesta in the middle of the day. Then it cruises until sundown and sleeps through the night. A fly follows this routine every day of its life even if it ecloses in a dark test tube all by itself and never sees a ray of light or another living being. Even if the fly's ancestors have been born and died in darkness for generations, like the citizens of Plato's cave, the fly still moves through its life in the dark at the same pace as the sun it never sees. Like fruit flies outside the laboratory or fruit flies in Casablanca, in Cairo, or in the Greek islands, it always takes that siesta in the middle of the day. Since it lives in a test tube in a pitch-black room, it does not need a midday break; but a fly in an open-air fruit market in Morocco needs a siesta to escape the heat, just like the human buyers and sellers in the market.

So waking and sleeping organize all of the animal's behavior, Hall said, and rising up and lying down are bona fide behavior patterns in themselves. "Not just 'cause I say so. It really *is* behavior. I'd argue with your friends for forty-five minutes, and at the end of it, I would demolish them."

Some people seem to think that behavior is behavior only when it is a mystery, Hall continued. But once any piece of behavior is understood at the molecular level, it all comes down to metabolism, whether we are talking about the way a weaver ant folds a leaf, a weaverbird weaves a hanging nest, a human being learns and speaks Swahili, or a fly rises with the dawn and settles down at dusk. "Benzer was once subjected, *in my earshot*," Hall said, "to some dumb question like 'Is that the mind or the brain?' But every aspect of mind and brain is ultimately metabolism! What do we think? Some kind of electric aura hovers around our heads?" We still seem to want something outside the mechanism, Hall said, some deus ex machina to save us from the clockwork that we have been exploring above and inside our heads for the last several centuries. It is now time for us to accept that behavior is as much a part of the material world as the stars above us and the atoms inside us. All behavior turns on molecular clockwork, Hall said, yet all behavior is fascinating.

Everyone who studies genes and behavior at the level of genes, molecules, or nerves gets asked questions like this: "Is *that* all there is?" Anatomists hear the same question when first-year medical students get their first look at the organs inside a cadaver. Behavior geneticists hear this all the time now, and the clock gene in particular attracts the question, Hall said, because a clock is a mechanism that we all understand to be a mechanism. It is the very symbol of mechanism in the heavens and in the body. So the deep assumption of the question, Hall said, is this: "Once you know something about it, it's not behavior. It remains behavior as long as it sits at the level of mystery and miracle." In Hall's opinion, that is why people ask him if keeping track of time is really a piece of behavior when it plainly is, and so is the act of seeing the light from a lightbulb, "some of the most fascinating behavior in the whole history of biological investigation!"

The reporter looked as if he were beginning to be glad he had said the question was his friends'. "Vision!" Hall shouted, waking the dogs under his desk. "It's not just *that you can see*! It's how you *respond* to the visual world of shapes and movements! And this is not to castigate your friend and say, 'You're an *idiot!*' This is not to say, 'Your ignorance is so colossal that you're doing nothing but sit there and dribble saliva *into your lap!*'" And Hall went through the roof. ("Jeff Hall easily goes through the roof," says Benzer—which is no more than Hall often says of himself.)

Other clock watchers answer the question in a lower key. Jerry Feldman, who discovered a clock gene in the fungus *Neurospora,* says calmly, "Call it what you want. You could say any gene that modifies behavior is a behavioral gene. What's important is the gene. What does it do? Ultimately, how does it couple with behavior? That's what interests me."

Kyriacou says cheerfully, "It's a real behavior gene. What it does, it adds behavior when it's there. It has rhythms. It's beautiful."

TODAY MANY of Benzer's students credit Hall with the success of the atomic theory of behavior after Benzer's beginning. "It's because of what Hall pushed first that the field came of age," says Tim Tully, one of Benzer's students' students.

In the late 1970s and for much of the 1980s, Hall felt embattled and alone, as if he were carrying a fallen banner. Benzer, with his fame, could have been a general not only for Konopka but for the whole field he had started. Benzer could have championed it like another Morgan. But unlike Morgan, Benzer had never had much time for committees or speeches, for writing books or playing scientific politics, and he was not working on behavior anymore.

In those years, Benzer was going more and more deeply into the nervous system of the fly. He had decided that before he went any further with behavior he should study the way genes build bodies. He and his latest students were using genes and mutants to study the way nerves work: doing basic neurogenetics. Tracing the way genes build the nerves in the embryo seemed to Benzer the next logical order of business.

Hall was outraged, but Benzer saw nothing wrong with jumping fields. He had jumped before. At Caltech during a night of skits after Delbrück had won his Nobel Prize, Delbrück's circle sang a song about Benzer to the tune of "Jimmy Crack Corn." In the first verses he was a physicist playing with charged particles:

> *Physics was fun, but I don't care,*
> *I'm on to something else next year,*
> *I must stick with the new frontier*
> *Until I'm old and gray.*

In the next verses he quit physics and mapped the gene:

> *Genetics was fun, but I don't care,*
> *I'm on to something else next year,*
> *I must stick with the new frontier*
> *Until I'm old and gray.*

In the last verses he quit genetics for behavior:

> *Behavior was fun, but I don't care,*
> *I'm on to something else next year,*
> *I must stick with the new frontier*
> *Until I'm old and gray.*

It was not only Benzer who was restless; by the late 1970s, many of the founders of the atomic theory of behavior had turned away from the theory. Sydney Brenner was phasing down his studies of worm behavior and studying the wiring diagrams of worm nerves instead. In fact, he had turned in this new direction before Benzer had, and his arguments had helped persuade Benzer that the embryo was the next frontier. Now at meetings, Brenner ragged Hall. "All of you neurogeneticists think you are finding interesting new things from your neural mutants," he told Hall, "whereas in the end they are all just going to be defective in things like aldolase," a garden-variety enzyme without which a fly cannot digest glucose and gradually starves to death. This was the same argument that Benzer had heard in the Sperry lab and Hall had heard from Michael Rosbash in the Brandeis locker room: *period, fruitless, dunce,* and the rest were just sick flies; there was something wrong with their house-keeping genes.

Another phage veteran who abandoned the atomic theory of behavior in those years was Gunther Stent, who had been working on the leech. Now Stent began writing pessimistic essays about the field, just as ten years before, in the brief doldrums of the 1960s, he had written elegiacal essays about molecular biology. Stent had now reconsidered Butler's saying "A hen is only an egg's way of making another egg." If life is a circle, why focus on the genes? Stent was beginning to look at life more in the philosophical spirit of Emerson: "The method of nature: who could ever analyze it? That rushing stream will not stop to be observed. We can never surprise nature in a corner; never find the end of a thread; never tell where to set the first stone. The bird hastens to lay her egg: the egg hastens to be a bird." Stent doubted that the genetic dissection of behavior or even the genetic dissection of development was going to turn out to be much more informative than dissection with a scalpel.

Meanwhile, Delbrück felt completely defeated. Twenty-five years after dictating his ebullient letter to Benzer about "starting a new life," Delbrück still could not understand how a stalk of fungus grows toward light. He confided to his diary that he was "sick at heart at the unsolved state of the problem." He had failed to understand the genetics and mechanics of even one simple-seeming piece of behavior. "I think Max indeed was disillusioned," Crick says now. "The phenomenon he tried to study was very intricate, and I don't think he ever got to the bottom of it. And it was, I think, not a wise choice in the first place, and a bit of bad luck as well."

In 1978, Delbrück's doctors, taking routine X rays, discovered that cancer was chewing away his ribs. Benzer and dozens of other phage veterans took to visiting him in his living room in the afternoons while he sat in his rocking chair quoting Samuel Beckett ("The light gleams an instant . . ."). Although Beckett and Delbrück had both won the Nobel

Prize in the same year, Delbrück had not met his hero in Stockholm ("No, he didn't come, the dog.").

At the same time, Benzer's wife, Dotty, was in the hospital with breast cancer. The two of them had been unusually close ever since Seymour was sixteen and she was twenty-one. She was a lark and he was an owl, but he always strolled into the laboratory in the afternoons holding hands with her. That was unusual behavior at Caltech, though it was a Fly Room tradition. Thomas and Lilian Morgan had worked side by side among the milk bottles as soon as their children were out of the house. Morgan's Raiders had named one of their female mutants *bobbed*. She had short bristles, and they named her after Morgan's technician, Phoebe Reed, who had just bobbed her hair. Sturtevant married Phoebe Reed.

Benzer had begun transforming himself into an expert on cancer biology and treatment from the moment that he and Dotty knew she was sick. Now he neglected the laboratory for long periods to attend national and international conferences on breast cancer. Even when he was at the lab, he spent hours on international phone calls to specialists in Switzerland; and even when he was satisfied that he was taking Dotty to see the best possible doctor, he would still arrive in the doctor's office with stacks of technical papers to explain and discuss. Martin Arrowsmith had lost Leora to a tragic illness; Benzer was determined not to lose Dotty.

That June, Delbrück gave the commencement address at Caltech. He spoke in the school's "Court of Man," flanked by buildings dedicated to behavioral biology on one side and to the humanities on the other. In a sense, the atomic theory of behavior is an attempt to unite those disciplines, behavioral biology and the humanities. In the long view, it aims to unite all of the sciences, all of philosophy, and all the arts. But Delbrück, speaking to Caltech's graduating seniors from a wheelchair, predicted that the union would never take place. "Indeed," he said, "we can take it for granted that science is intrinsically incapable of coping with the recurrent questions of death, love, moral decision, greed, anger, aggression." And he told a weary fable. Science is like Tithonus of the Homeric myth, he said. When Tithonus was young, the goddess of the dawn, Aurora, fell in love with him. She asked Zeus to make him immortal. Zeus did so—but he did not grant him eternal youth. "Tithonus aged and shriveled and talked incessantly," Delbrück said. At last he turned into a grasshopper, and Aurora had to keep him in a little box.

"Science does chatter and chirp incessantly," Delbrück concluded, "sweet music to those few who are tuned in to it, but does it satisfy

Aurora's yearnings, Aurora the morning dawn?" Does it answer the questions we all want answered?

Benzer's current crew of postdocs sometimes talks about the history of the laboratory during their brief late-morning cigarette breaks, before Benzer comes in. They sit on stone benches under palm trees and jacarandas outside Church Hall, and they pass on the legends they have heard. For them all of this happened in hearsay time. They say that when Max and Dotty died, Benzer's friends doubted that he would ever recover. They say that for a while he published nothing, as if what he saw through a microscope was intrinsically incapable of touching the recurrent questions.

# CHAPTER THIRTEEN

---

# Reading an Instinct

*I am a book I neither wrote nor read.*

—DELMORE SCHWARTZ

MOST OF THE FOUNDERS of the gene business left their laboratories sooner or later to run corporations, foundations, universities. Crick served briefly as president of the Salk Institute in La Jolla. Watson became director of Cold Spring Harbor. "And that is apt to happen to people," says Crick in the cheerfully sturdy tones of Odysseus describing the whirlpool Charybdis. "It's not likely to happen to Seymour, who is the last person to want to do administration, I would have thought."

As Benzer's friends succumbed one by one, Benzer presented each of them with a copy of a little handbook called *Microcosmographia Academica: Being a Guide for the Young Academic Politician.* The pamphlet ("the merest sketch of the little world that now lies before you," the preface explains) teaches the skills that Benzer's former labmates would now need to learn, including propaganda, "that branch of the art of lying which consists in very nearly deceiving your friends without quite deceiving your enemies."

Benzer, of course, had always preferred solitary work, eased by the company of other night owls in the lab and by his wife and daughters at home. Although he was happy in the old phage days and the first mapping days, there had been moments when the work got too solitary, even for him. Back in the fall of 1952, when he returned from the Institut Pasteur to his lab at Purdue, he had written a lonely letter to Max Delbrück: "After Paris, this local phage isolation is almost unbearable."

By the mid-1980s, however, the map work that had begun in the night thoughts of Sturtevant and Benzer and a few others was exploding into big science: the biggest single project in the history of biology. One of its formative meetings took place on a weekend in May 1985, when the molecular biologist Robert Sinsheimer, now chancellor of the University of California at Santa Cruz, called in a few other molecular biologists to talk about the idea. UCSC had hoped to build a giant telescope, but the school had lost that project to Caltech. Now Sinsheimer was swiveling his sights around—like Galileo, who had turned his first celebrated telescope into a microscope and stared at a fly in a bottle. Sinsheimer was thinking about a project that he found even more exciting than a big telescope: a group effort to map every single gene in a human being.

This enterprise got away from UCSC too. The Human Genome Project was soon a multibillion-dollar public program with a vast bureaucratic apparatus involving the U.S. Department of Energy (DOE), the U.S. National Human Genome Research Institute (NHGR), for the coordination of international collaboration the Human Genome Organization (HUGO), and many more agencies and acronyms. Laboratories across the United States, France, Italy, the United Kingdom, and Japan began half racing and half collaborating to map and sequence every one of the 3 billion letters in the human genome across 3,600 map units. What Benzer did by hand in individual petri dishes was soon being done day and night by robots that sprayed human DNA into rows of vials; copied each snippet of DNA; chopped it up; sequenced the letters; and stored the letters in computers. By the last years of the twentieth century, the Institute for Genomic Research in Rockville, Maryland, would be sequencing millions of letters of code a year and hoping to go faster. The Washington University Genome Sequencing Center in St. Louis would be sequencing about 100,000 letters a day and hoping to go faster. The map will eventually fill the equivalent of 134 complete sets of the volumes that Sturtevant loved to commit to memory in the evenings, the *Encyclopaedia Britannica*. Biologists would compete to see how many letters of the human genome their "high-throughput" robots could decipher in how little time in laboratories they called factories. They talked patents and quality control and venture capital and data isolation (meaning secrecy).

James Watson became the first director of the Human Genome Project. He worked the telephone from Cold Spring Harbor in a voice of old celebrity, manipulating the multibillion-dollar government program and the needs of the pharmaceutical companies and biotechnology companies that eventually began taking a bottom-line interest in it. Wall Street trusted that the rational design of drugs would be radically enhanced by knowledge of genes and how they work. The Concorde and the first-class cabins of Boeing 747s were already beginning to fill with instantaires, scientists who had made a killing overnight in computer hardware and software or genetic hardware and software. Entrepreneurs raced through the last years of the century windmilling their arms: "I just sold one hundred thousand genes to SmithKline Beecham!" They were rich men who got to heaven through the eye of a needle—or, in the Arrowsmith view of life, fell from grace. Recently the star molecular biologist at Genset in Paris explained to a reporter from *Science* magazine why he plays the piano in the evenings. "In genetics there is no mystery," he said, "but music is all mystery."

Today in the president's office at Cold Spring Harbor, Watson deals with the acronyms along with the minutiae of Cold Spring Harbor hirings and firings and high-level dinner parties, conversations in which grungy young drosophilists stopping by from the Fly Room can hear the expensive thunk of limousine doors. "I might be going. . . . It's a question of if I can see Bill Gates." At luncheons Watson talks stocks and bonds and options. Cold Spring Harbor was the scientific center of the eugenics movement in the first half of this century. As the laboratory's director, Watson has helped preserve its eugenic archives as cautionary tales. But he likes to entertain his luncheon partners, the presidents of nations and corporations, with war stories of the bureaucratic or bio-cratic life, and he likes to strike a cynical tone. When Watson talks about the Human Genome Project's decision to set aside money for ethics research, for instance, he explains his reasoning with a smile: "To preempt the critics." At first, ethicists were making the birth of the program a very difficult labor. So Watson allotted them money. "Ethicists are a mixed lot," he says, "generally not worrying about their own problems, just somebody else's problems." He seems to speak with confidence, as if he were absolutely sure of approval.

Watson saw the Human Genome Project as the logical culmination of a career that began with the double helix, and as with the double helix he did everything he could to make it happen. "There were some specialists who were against it. So we bribed them," he likes to say with the same smile, explaining the Human Genome Project's early decision to sequence the genomes of a fruit fly, a nematode worm, a mouse, the bacterium *E. coli,* and the mustard weed *Arabidopsis.* But this was not really a bribe. Molecular biologists needed all of these model organisms and more if the code of human beings was to mean anything more at the end of the day than a series of Cs, As, Ts, and Gs. By sequencing the DNA of other species along with the DNA in their own bodies, the biologists would produce a kind of Rosetta stone. They would generate enough different versions of enough different genes to figure out what most of the words mean. It had become not only a question of science but also a question of high finance to figure out the meaning of genes—or even to figure out how to figure out the meaning, a difficult question with writing as ancient and peculiar as this. New companies sprang up just to elucidate the genes in what are usually called (in spite of Benzer's protests at meetings) the lower organisms.

SO IT HAPPENED that in the 1980s the work that Benzer and his students had started in middle-of-the-night eccentricity and fly-on-the-wall obscurity began heating up. Observers in and around biology were beginning to realize the value of flies and neurogenetics. Jeff Hall thought that *Drosophila* was becoming industrial. At conferences on *Drosophila* neurobiology, which were now legion, there were ten scientists competing for every speaking slot. For a while the study of genes and behavior remained a fringe specialty. Most of the great names in neurogenetics did not work on behavior. Instead, following Benzer's lead, they studied the way fly nerves grow in embryos, the way fly nerves speak to one another in the brain. At fly meetings, behavior was still the caboose of the train. The last session was always the behavioral session, which the chairman of that session had labored to fill because there was still hardly anybody working in the field. Hall often spoke to an audience that, in effect, consisted of the janitor who was cleaning out the auditorium. He says that even neurogeneticists thought he was working with "silly and embarrassing behavioral observations, and titillating mutants."

"This sounds almost like whining, but it isn't." Hall preferred problems that no one else cared about. It would have cost him his soul to jockey with the journeymen in the field of nerve growth. "That's so solid," he used to tell people. "It's blood and guts. Whereas behavior just seems this vast, open-ended chaos. In other areas you can just fight your way through with more work, and you get through the molecular or anatomical problems you face. But with behavior, half the time it just seems we're just floundering and we don't know how the hell to cope with it. So it's not a complaint at all. I don't work on behavior and feel an outcast." Hall felt like an outcast, so he worked on behavior.

Benzer hated the new crowds too. His laboratory's work on the growth of fly embryos, fly nerves, and in particular fly eyes had made the fly eye

one of the hottest fields in neurobiology. "And there was enormous competition between Seymour's lab and Gerry Rubin's lab in Berkeley,"
recalls Michael Ashburner. "That's probably passed now, because they've
each moved on. But at one time they were both doing rather similar
experiments and certainly following a rather similar strategy, and I think
there's no question that Gerry's lab did it more successfully than Seymour's—that the tension of that competition wasn't helpful. This is just
an anecdote. But I was at Berkeley in '84, '85, can't remember now. And I
remember going down to Caltech. Fifteen or sixteen people sat around
in Seymour's coffee room, had sandwiches, and chewed the cud for kind
of an hour and a half. And eventually Seymour said, 'Well, back to work,
chaps.' And I said, 'Well, I've done my bit for Gerry Rubin.' He was,
'What do you mean?' I said, 'Well, I've kept all your lab off the bench for
the last hour and a half, didn't I?' " Ashburner cackles. "He turned absolutely white. He didn't think it was funny at all. It was true, of course."

Now Konopka's gene *period* was no longer an object of merely eccentric or philosophical or human curiosity. Such a gene had to be connected in some unknown but intricate way with the living machinery of
the fly. So *period* became a model for the way nested cogwheels of genes
work together to produce behavior, and more fundamentally as a model
for the way genes work and for the way to figure out how genes work.
Serious science and money were riding on these questions. The two
principal molecular biologists among the clock watchers, Rosbash and
Young, each won grants from the Howard Hughes Medical Institute
guaranteeing their laboratories funding of $1 million per year. Hall did
not win a Hughes. "I'm a pseudo–molecular biologist," he says, "a pseudo-
crypto–molecular biologist. I'm really more of a geneticist." He had
always been closer to Morgan's tradition, and now he felt more an outcast than ever. Konopka was in even worse shape: he had failed to get
tenure even at Clarkson.

With the new tools of molecular biology, the standard order of business is cloning and sequencing. Molecular biologists clone a gene, make
billions of copies of it, chop the gene into bits, and find out how the gene
is spelled—for instance, GCTAAAT . . . —the same routine to which
the robots of the Human Genome Project are dedicated. To find out
what a given gene's sequence means, molecular biologists can then alter
its spelling one letter at a time, changing a G into an A or an A into a C.
They can clone each of these variant genes, insert each variant into a fly
embryo, and watch to see how each change in spelling changes the
gene's behavior. By cloning a gene from a human being and inserting it

into a fly, a worm, or a mouse, or by changing the spelling of a fly gene and reinserting it into a fly, molecular biologists can get clues about the nature of the gene.

In this way, every week, another old and unanswerable question becomes answerable. Recently a team cloned and sequenced a gene that makes Mendel's peas wrinkled or smooth. The difference lies in a tiny fragment of DNA in the middle of the gene, a little block of letters that has jumped into the gene: the kind of jumping gene that McClintock studied in maize. These jumping genes are now known as "transposons." This particular transposon is stuck in the pea gene like the proverbial lost wrench in the engine. If a pea plant inherits a version of the gene that has the transposon stuck in it, the plant cannot make a certain enzyme, starch-branching enzyme I. Without that enzyme, the plant cannot make a certain protein, branched-chain amylopectin. Without branched-chain amylopectin, the pea plant's cells tend to fill up with abnormal amounts of sugar. The peas swell. When they dry out, they wrinkle.

Molecular biologists have also cloned and sequenced a gene that makes Mendel's pea plants tall or short. That gene codes for an enzyme that manufactures some of the building blocks of a plant hormone known as GA1. GA1 is a gibberellin, a growth hormone. One version of this gene makes an enzyme that manufactures the building blocks with a single amino acid misspelled. Because of that single misspelling, the pea plant produces only one twentieth the normal quantity of growth hormone, so the plant grows up short.

By the late 1980s, in principle, anyone who wanted to know why *white* has white eyes could now find out. Anyone who wanted to understand the workings of the strange genes that Benzer and his raiders had discovered could find out. The original mutants had been novelties, and it had been extremely difficult and probably impossible to work out the function of many of them simply from the classical genetic analysis that Benzer and his students had done in his Fly Room. Merely mapping the genes to their chromosomes, as Benzer and his crew had done, using the tools of classical genetics, would never have told them how the genes work. To understand that, they had to find out what protein each gene makes. Then they had to track that protein through the body and see what the protein did. Even today that kind of search can be difficult, but in the early days of Benzer's Fly Room it was impossible.

.    .    .

IN 1986—once again in a dead heat—the rival clock laboratories in Boston and New York determined the complete sequence of letters in the *period* gene. The portion of DNA that encodes the *period* protein is a length of approximately 3,600 letters of genetic code. No one had ever sequenced a gene that shapes behavior. Now that they had the sequence in front of them, the neurogeneticists could see exactly what is different about *period zero,* the first mutant that Konopka had put on the map, the mutant with zero sense of time. The mutation that makes its behavior so eccentric is located in the first half of the *period* gene, which is now known familiarly as *per.* At nucleotide 1390, counting from the start of the coding sequence, the letter C has changed to the letter T. This point mutation transforms the three-letter word CAG (which means "glutamine") into the three-letter word TAG (which means "stop"). So when a fly cell comes to that TAG, the cell's production of *period* protein stops. At that point the cell is close to finished with its production. It has already transcribed its *period* DNA into *period* RNA. Ribonucleic acid (RNA) is the compound that carries genetic messages from the DNA inside the cell's nucleus to the protein-building machinery outside the nucleus. But the cell's translation of *period* RNA hits that stop about one third of the way through its production of the protein, and the fly can never finish the job. Over and over the fly manufactures that same useless fragment, like a worker on an assembly line who has received a torn sheet of instructions.

The clock watchers could also see what is wrong with the code of Konopka's *per short* and *per long,* the mutants with clocks that run five hours faster and five hours slower than normal. In *per short,* at nucleotide number 1766, the letter G has changed to the letter A. In *per long,* at nucleotide number 734, the letter T has changed to the letter A. These changes do not break the clock, they only accelerate or decelerate the hands. Again a point mutation, the smallest possible mutation, makes the difference, producing short days or long days for the fly, just as the change of one letter of code gives short stems or tall stems to peas.

Hall, Rosbash, Young, and their teams were now in an unprecedented position to trace the mechanical connections between a gene sequence and a piece of behavior. They had opened the back of the clock, and now they could try to feel their way through the mechanism from gene to metabolism to behavior—which had been Benzer's original goal for the work when he had founded the field. For this enterprise a clockwork gene was the perfect place to start. The internal workings of the system could safely be assumed to be extremely regular. And variations in the behavior stood out clearly because the behavior they were studying runs like clockwork too. "Some fly behaviors are just chaos," Hall says, "hopelessly irreproducible, a total mess. But rhythms are not that way." So the clock gene made an ideal model, a perfect point of entry into the connection between genes and behavior.

Now that they had cloned and sequenced the gene and discovered that spurious stop, they translated the rest of the code, triplet by triplet, which told them every amino acid building block in the *period* protein. They had hoped that finding the protein would lead them straight into the clockworks. But Hall, Rosbash, and Young soon discovered that in spite of all they had learned so far, they were still stymied because they had no idea what the *period* protein actually did in the fly; nothing quite like it had ever been seen before.

"Because we are at the beginning of something, we don't know what is going to emerge," Rosbash used to say stoically. "It's like a garden in spring. Very *early* spring. You know, like, *winter.*" It was a long, cruel season of search and blunder. For several years they struggled with no forward motion. All of the principal investigators suffered errors of judgment that make them blush now ("Partly a function of hysterical racing, of racing other people to get the answer," says Rosbash). For a time they thought the *period* gene might be involved in the way certain cells in the brain communicate with one another; they thought the *period* protein might work in the gap junctions between nerves to set rhythmic pulses flowing through the brain. And for a time they thought the *period* protein was a proteoglycan, which is a protein as brambly, tangled, and hard to work with as a thornbush in a bog.

"This phase of the story, from '85 to '87, was the low ebb of work on this *or any other* gene," Jeff Hall says today. He, Rosbash, Young, and Kyriacou often felt as if they were stuck in a sand trap in the middle of an hourglass. "And outsiders said the same thing," Hall remembers. " 'You guys make all these crazy findings and claims, and they don't seem to make any sense. What is true? What is going on here? Does *anything* make sense? Are *any* of the simple data of *any* validity?' "

It was the archetypal project of the new biology. Where the first decades of the century merely mapped gene to trait, they were trying to go into the clockwork and trace it all the way from the gene to the movement of the hands on the clock. The rival clock labs were finding out just how difficult that is; and they were not alone. Elsewhere, progress in the study of hundreds of other genes was also agonizingly slow. Having found the approximate location of a gene on the map of the chromosomes, investigators thought they would soon be able to clone it, sequence it, and find out what it does. Then they would have the first clues to the causes of diseases and the way to treatment. That research model was the source of the power, money, and people flooding into molecular biology. Investors thought finding genes was everything. They were encouraged in that belief by able propagandists such as Watson, who told *Time* magazine in 1989, "We used to think our fate was in the stars. Now we know, in large measure, our fate is in our genes." Like Francis Galton before them, the leaders of the field sometimes spoke now as if the next sunrise could banish what the ancients had understood so well: the twists, turns, and depths of human nature. The dreams they dreamed or encouraged others to dream showed an insufficient respect for the complexities of human unhappiness and for the depths of physiology that lie between gene and behavior. Meanwhile, down in the trenches, molecular biologists were beginning to understand that even after they found, mapped, cloned, and sequenced a gene, their search had only begun. They still could not do anything with the gene until they knew what the gene does. Every gene is a thread that leads into vast skeins of molecular anatomy, and one by one molecular biologists have discovered how easy it is to get lost at the very beginning of the thread. There is so much they do not know.

Konopka put the *period* gene on the map in 1971; Hall and his friends and rivals sequenced it in 1986 and 1987; and even in the early 1990s, no one had a clue how the clock worked. Drosophilists and drosophilosophers gossiped about the *period* problem in Fly Rooms around the world: "Ah, yes, *per*. The gene that promised so much and never quite delivered."

# CHAPTER FOURTEEN

# Singed Wings

*Philosophy is really Homesickness; the wish to be every-
where at home.*

—NOVALIS

E. O. WILSON had published his famous book *Sociobiology: The New
Synthesis* in 1975. There he analyzed the social instincts that bring
together colonies of ants and bees, herds of wildebeests and antelope,
tribes of chimpanzees, pairs of flies. In the last chapter he turned to
human beings and argued that we are instinctively social animals too.

That fall, the molecular biologist Jonathan Beckwith, the population geneticist Richard Lewontin, and other colleagues of Wilson's
denounced sociobiology in highly publicized attacks. They argued that
Wilson could not make this leap; that although the rest of the animal
kingdom is shaped by instincts, the human animal has gotten free of the
slavery of instincts and works largely by different rules, the rules of culture. It was the nature-versus-nurture argument of the first half of the
twentieth century in its most concentrated form. Wilson was vilified by
bullhorn in Harvard Square, his classes were leafleted and picketed, and
there were times when he was afraid he would be physically attacked by
demonstrators, although the worst that ever happened was when a
woman at a scientific meeting in Washington dumped a bucket of ice
water over his head while protesters from a group called the International Committee Against Racism chanted, "Wilson, you're all wet."

"I had unexpectedly—unexpectedly to me, anyway—stumbled into a
hornet's nest of resentments and fears that represented, to some extent,
a holdover from the activist period of the sixties," Wilson says now.
"When I got into this controversy, I realized early on that I had both a

challenge and a responsibility to explore the subject further." So Wilson did more thinking and reading in 1975, '76, and '77, and wrote the book *On Human Nature*, which won the 1979 Pulitzer Prize. There he explored some of the earliest work on the language instinct, homosexuality, and other topics in human behavior genetics, and tried to interweave the science with philosophical and ethical considerations. Today many of his case studies seem dated and speculative, but the book's basic argument is straightforward. When we look at ourselves against the background of the other animals, we can begin to see that we, like they, have distinctive traits. These traits most resemble those of our nearest living relatives, the chimpanzees. These facts support the hypothesis that we are shaped by our genes; they contradict the hypothesis that we have escaped our genes. It is hard now to appreciate the heat that this thesis could engender at that time and in that place. "It's died down a great deal since the late seventies," Wilson says. "It was worth your head to discuss these subjects too openly in the seventies."

In the 1970s, Benzer and his school were studying genes and behavior too, and they were studying the subject from the one angle of attack through which knowledge becomes power. But they worked so far outside the hot glare of lights that lit and singed Wilson that histories of the controversy do not even include Benzer in their indexes. He never attracted Wilson's enemies, partly because he never wrote a book. He was also apolitical, like his mentor Delbrück before him. On the day of his Nobel Prize ceremony, in 1969, Delbrück had written in his diary that he was depressed: "The main reason for my depression is my feeling guilty. All the time one is questioned about items one doesn't know anything about, though one should. These questions refer to a world outside the ivory tower which I used to ignore successfully."

The outside world ignored Benzer successfully. Yet if any research program in biology deserved to be watched, his did. He and his students were dissecting genes that are central to animal behavior: points of entry into time, love, memory. "But that took a while to *prove*," says one of Benzer's students' students, Ralph Greenspan, who now runs a Fly Room at the Neurosciences Institute in San Diego. "And between the time when you report and the time when you prove . . ." Until quite recently there were grounds for skepticism. Benzer and his students were also protected from serious attention by their laboratory animal. Today Benzer sometimes exclaims when he sees a headline, "We were doing that thirty years ago! The fly is always regarded as kind of an abstraction."

In a quiet way, the molecular biologists were doing work that would eventually help to vindicate Wilson and some of the aims of sociobiology, long after the subject had been so raked over and tarnished that virtually everyone in the field had abandoned Wilson's name for it. "Wilson always hated the molecular biologists," says one of his colleagues at Harvard's Museum of Comparative Zoology. "And yet they're the ones who won his war."

The war did singe Benzer too, just as he was turning away from behavior. In 1979, he was asked to deliver a lecture about genes, learning, and memory at a plenary session of the Sixteenth International Ethological Conference in Vancouver. That September, a few weeks after the lecture, he received by certified mail a six-page, single-spaced letter attacking his work. The same letter appeared in the mailboxes of the entire Caltech faculty and all of the invitees of the Vancouver conference.

The letter came from Jerry Hirsch, a behavior geneticist at the University of Illinois in Urbana. Hirsch was a strong polemicist against eugenics and racism. He had learned from one of Morgan's Raiders, Theodosius Dobzhansky, to think of genetics as a powerful argument against prejudice of all kinds. Hirsch had once published a manifesto on the subject in *Science,* back in 1963, while Benzer was still feeling his way from *rII* toward behavior. In fact, in Hirsch's manifesto, "Behavior Genetics and Individuality Understood," he had used Benzer's *rII* work to calculate that the probability that two parents will have a second child genetically identical to their first is less than one in 70 trillion. "Individual differences are no accident," Hirsch argued, and every individual is so distinctive that racism is intellectually bankrupt: "The concept of a normal individual has no generality." Hirsch had spent years breeding fruit flies for changes in their instinct to turn upward rather than downward in a vertical maze: geotaxis. He ran his flies through the maze individually, and he hoped that by dramatizing the genetic diversity that underlies behavior, his experiment would strike another blow against racism.

Benzer's countercurrent machine had been inspired in part by Hirsch's vertical labyrinth. But Benzer was looking for mutants that might provide a point of entry into the molecular basis of an instinct. In this way Benzer was working toward a goal that was out of reach for Hirsch. Traditional behavior geneticists like Hirsch worked with the tools of yet another branch of the science that Mendel and Morgan started, a branch known as population genetics, which is strongly mathematical. After a statistical analysis of his breeding study, Hirsch could not map the genes involved in an instinct like geotaxis. He could conclude, for instance, that no more than two or three genes seemed to make the difference. But he could not find the genes themselves. He had to "declare victory and retreat," as one of Benzer's students' students once put it, whereas Benzer could find single genes and then map them, hoping eventually to use them to split behavior patterns into pieces, dissect them, and see how they work.

In the Vancouver letter, Hirsch criticized Benzer for herding squads of flies through his machine. Benzer was behaving as if the flies were genetically homogeneous, except for a few oddball mutants. In this way Benzer was committing what Hirsch often called "the uniformity assumption," the original sin not only of many failed behavior studies but also of racism.

Hirsch thought his disagreements with Benzer ran deep, although Benzer had trouble understanding most of them. Hirsch had also spoken at the Vancouver meeting. "Hirsch traditionally gave some kind of bigwig talk," says a former student of Hirsch, Tim Tully, who later went over to the Benzer camp. "This was *his meeting,* OK?" But the conference organizers had asked Benzer, not Hirsch, to give the plenary address. Tully, who was an undergraduate student at the University of Illinois a few years before the episode, remembers Hirsch pulling him into his office once in 1973, saying, "I've got it, I've got it." Hirsch showed Tully a *Scientific American* article by Benzer, "Genetic Dissection of Behavior." Hirsch read him Benzer's whimsical description of a one-eyed fly that makes a helix as it runs up a tube toward a light. Hirsch told Tully—with what Tully remembers as an unnerving intensity—that Benzer had written "clockwise" when he should have written "counterclockwise." Tully walked away, mulled over Hirsch's argument, and realized that Benzer was right. Tully still remembers the moment he broke the news to his professor. "He was about to launch another letter. He stood up and closed the door in my face."

"I really didn't understand," Benzer says now. "I thought it was way off base, trying to ruin my reputation with my colleagues. When somebody denounces you like that, everyone thinks, quite aside from the facts of the matter, 'Well, you know . . .' There's always this doubt: 'Maybe the guy has something,' " Benzer says. He sensed that doubt in his colleagues whenever he tried to talk with them about the letter. Eventually he decided to give a department seminar to clear the air. The talk was standing room only. It was not a success. "I don't think I did a very good job," Benzer says now. "I tried to explain everything from both sides, and it did not go over very well."

Konrad Lorenz liked to describe the territorial disputes of male stick-lebacks defending their nests. With sticklebacks as with academics, a property dispute is a ritualized head-to-head dance that seldom draws actual blood. "The pursuit is repeated a few times in alternating direc-tions," Lorenz wrote, "swinging to and fro like a pendulum which at last reaches a state of equilibrium at a certain point." By sending such a pub-lic letter, Hirsch was going head-to-head with Benzer, and the battle might have been expected to swing back and forth for years. But after that one failed seminar, Benzer did not mount a counterchase. Instead he let the learning-and-memory work go, along with all the rest of his mutant behavior work. Although he does not say so, it was a bad time for him, the year after Dotty's death. His two daughters were grown and gone. For the first time in his life, he was coming home from the lab at dawn to empty rooms.

"Well, people said, 'This man just craves attention,'" Benzer says today. "Roger Sperry told me, 'Don't give him attention, or there'll be no end to it.' So, I just gradually forgot about it. A memory like that decays rather slowly. Now I'm glad that my reputation has survived somehow."

For years, Hirsch would continue to argue that Benzer and his stu-dents were hopelessly misguided, even after their discoveries began appearing almost monthly on the covers of *Science, Nature, Cell,* and *Neuron.* When asked about the science that Benzer had started, Hirsch would say, "It's certainly moving. Maybe not forward."

And for years, Benzer would keep a distance from the atomic theory of behavior. He decided to make a fresh start, as he had done so many times before. When he turned away from electronics, his interest had shut off like a tap. He had turned away from gene mapping in the same way, and after Dotty died he could not bear to think about cancer.

In his sanctum at Church Hall, Benzer lost himself in the microcosm in which he felt increasingly at home. Through the microscope he stud-ied the eye of the fly. Each fly eye has eight hundred facets called "ommatidia." Each ommatidium is a hexagon, and at high power he could see inside each hexagon a bundle of eight cells in the shape of a trapezoid, with six cells on the outside, R1 through R6, and two on the inside, R7 and R8.

Again and again he watched the living cells of a fly embryo build this structure. Again and again he watched the eye grow as if it were the latticework of a living crystal, a neurocrystal. He could actually see the nerve cells' nuclei bobbing up and down inside them as they chose their places and decided their fates one by one inside the growing dome.

Benzer had imagined that to build such an elaborate neurocrystal, the genes and cells must work like cogwheels, like clockwork, like atoms falling into place in inorganic crystals with absolute regularity. But even in the eye of the fly, he was amazed to see the nerve cells making decisions and revisions based on where they were at each particular moment and what was happening around them. It was almost like watching the flies in his countercurrent machine. And here, too, there were mutants. A mutant fly called *sevenless* was discovered by two of Benzer's graduate students, William Harris and Donald Ready (who devoted his Ph.D. work to the eye of the fly). In *sevenless,* a nerve cell that by rights should become R7 strays off instead to another fate. Instead of a photoreceptor cell, it becomes a cell that cannot detect light, and it helps make the lens of the ommatidium. In other words, R7 behaves very much like a fly that turns toward the darkness instead of the light. A single mutation makes the difference.

WHILE BENZER was exploring the eye of the fly, he met Carol Miller, a neuropathologist who worked across town at the University of Southern California School of Medicine. She was almost twenty years younger, and Benzer thought she looked like the woman in a painting he loved, Vermeer's *Woman with a Turban.* They soon discovered that they shared an obsession with the machinery of the brain. They both liked the same kinds of novelties and gewgaws, too: a plastic brain with a compass in it; a plastic eyeball key chain; a brain-shaped Jell-O mold. They both liked the same books: *Human Oddities; Smith's Recognizable Patterns of Human Malformation;* and *Sideshow,* which has photographs of dwarfs, midgets, giants, women with beards, a woman with a breast on her thigh. He was dissecting fly eyes and brains; she was dissecting human eyes and brains. They could talk about their work for hours over a Friday-night dinner of sushi, squid, or *cervelles de veau en matelote.* Seymour would tell Carol about the bizarre mutant fly he had found that week, and Carol would say, "That sounds like the patient I saw yesterday."

Seymour and Carol got married, and in the early 1980s they embarked on a joint research project. In his Fly Room, one of his postdocs, Shinobu Fujita, had been taking bits of fly brains and injecting them into a

*Benzer watched the eye grow as if it were the lattices of a living crystal. A single fly eye has eight hundred hexagonal facets.*

mouse, just as if he were vaccinating the mouse against measles or chicken pox. The mouse would make antibodies against the foreign fly proteins. Then Fujita would clone those antibodies one by one, make a big batch of each, and use it to stain fly brains. With these monoclonal antibodies, as they are called, the Benzer group could stain the fly eye and brain with exquisite specificity. That is, each antibody attached itself to a specific fly protein or to a specific lump or angle of one protein and no other.

For their joint project, Carol and Seymour decided to test these antibodies on samples of the human central nervous system. Since antibodies are so specific, Seymour thought it unlikely that they would stain anything human. But it was a whimsical adventure to start their marriage: a marriage of true minds, as Carol told her friends. She took postmortem samples from four young human patients, cutting little blocks of about one cubic centimeter each from the spinal cord, optic nerve, hippocampus, cerebellum, lymph node, and liver. She deep-froze these blocks and prepared them in much the same way that Seymour had prepared his flies, shaving each frozen block very fine to make a series of microscope slides and applying the antibodies to each slide, one by one. If her patients had proteins or bits of proteins that were exactly like flies' proteins, the stain would light them up in bright green.

When Carol looked through her microscope, she was startled to see bright green on slide after slide. Almost half of the antibodies stained her human specimens. She found hit after hit. She and Seymour sat together in her laboratory, looking at the slides and shaking their heads. In effect, the antibodies had scanned the fly brain and the human brain and found dozens of identical molecular building blocks, which implied dozens of identical gene sequences. When Benzer had opened his Fly Room, he had hoped that the fly might have something to say about human genes and behavior, at least by way of analogy. But he had never dreamed that the fly brain and the human brain would turn out to have this much molecular machinery in common.

Carol was just as surprised. The family likenesses were odd. One of Seymour's antibodies stained just a small and specific portion of the retina in the human eye and the equivalent portion of the retina in the eye of the fly. Another antibody stained only the Purkinje cells in the middle layer of the cerebellum. Purkinje cells are the most spectacularly dendritic trees in the human brain. Each Purkinje cell has so many tiny branches and twigs that it makes 150,000 contacts with nerves around it. In other words, a single one of these cells makes about as many contacts in the human brain as a fly has brain cells. Many neurobiologists believe that this prodigious branching of connections produces the human brain's unprecedented power. Biologists who hate sociobiology maintain that we can learn very little about human nature by studying the instincts of other animals because the powers conferred by interdigitating neurons like these lift us too far beyond the other animals for their instincts to illuminate ours. In Carol's slides, however, Seymour's antibodies circled each and every Purkinje cell with dotted lines like broken halos. So although our brains are fancier, they are made of the same stuff.

For Carol the stains were a bonanza. She saw immediately that they would help her explore the human nervous system, both normal and pathological, in finer detail than before.

For Seymour, sitting in Carol's pathology lab surrounded by jars of gray-brown, half-dissected human brains and gray-white human eyeballs, the green stains meant something else. They meant that at the

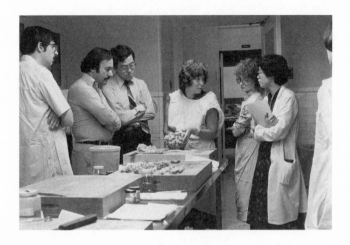

*A marriage of true minds. When Seymour Benzer met the neuropatholo-
gist Carol Miller, they began discovering uncanny connections between
the genes of flies and human beings. Here, Carol holds a human brain in
her laboratory at the University of Southern California School of
Medicine, in Los Angeles.*

level of genes, the family resemblances between species at far distant
reaches of the tree of life must be very close. His old mutants might turn
out to be more interesting than he had thought.

THE UNIVERSAL FAMILY likeness of genes was soon confirmed again and
again around the world as the first sequence data began to come in and
were published in computerized databases. Biologists typing in a search
for a string of genetic code they had just sequenced in a fly found hits in
the ox, the goat, the sculpin, and the mushroom. It was a moment of
huge collective simultaneous discovery. There has never been anything
quite like it in the history of science. Tens of thousands of genes or gene
sequences in human beings turned out to have close counterparts in the
fly, the worm, the yeast, the mustard weed, and even *E. coli*. The discovery would have thrilled Darwin, although it was an old truth for poets.
George Herbert wrote in "Man," "Herbes gladly cure our flesh; because
that they/Find their acquaintance there."

The news caused at least as much excitement on Wall Street as it did
in the academies. In Cambridge, England, where the double helix had
been discovered at midcentury, molecular biologists founded the company Hexagen to look for clues to human diseases in the genes of the

mouse. In Cambridge, Massachusetts, molecular biologists founded NemaPharm to look at the genes of the worm and Exelexis to look at the fly. The chief drosophilist at Exelexis explained to investors recently that it is astonishing how much one can learn from flies at the level of the genes: until the mapping projects began, he said, drosophilists had never realized "that we were looking at little people with wings."

This was one of the great developments of late-twentieth-century biology. It means that every biologist studying every genome can now feel more or less at home in all the rest. And Seymour cherishes a sentimental feeling for the way he and Carol got their private preview of the family likenesses in the tree of life.

To this day, Carol and the pathologists in her laboratory are using Seymour's antibodies to explore neurodegenerative diseases of the human brain, such as Alzheimer's, Huntington's chorea, Parkinson's, and amyotrophic lateral sclerosis (Lou Gehrig's disease), a disease her mother developed. Carol and Seymour cared for her in their home in San Marino. The disease killed her mother, but Carol kept tissue samples and blood samples for her research program. Carol's strongest interest was in Alzheimer's disease, and she found abnormalities in vulnerable neurons that no one had ever described. Today Seymour's fly strains are helping Carol untangle the tangle of Alzheimer's by tracing precisely what goes wrong in the populations and subpopulations of cells that Alzheimer's slowly blights.

Meanwhile in Benzer's laboratory his Korean postdoc Kyung-Tai Min is screening fly mutants, searching for brain degeneration patterns that literally look like human ones. That is, Tai is trying to find fly brains and human brains whose problems look alike by visual inspection. First, he screens for mutant flies with reduced life spans. Then he dissects their brains and examines them with an electron microscope. He is finding all kinds of distinctive lesions in the flies' brains that look remarkably like the lesions in the textbooks that Carol has lent him from her pathology lab: pictures of brains with Alzheimer's, Parkinson's, Lou Gehrig's disease, Huntington's. When Tai finds a particularly interesting case, in which both the symptoms and the brain stains are alike, Seymour tells Carol about the case over dinner before he goes back to the lab.

His old hero Arrowsmith also remarried—though not as happily. The front door of their house in San Marino has a stained-glass window they designed together in which fly brains and fly eyes are decoratively disguised as flowers. A framed reproduction of Vermeer's *Woman with a Turban* hangs in the den.

Sometimes Carol and Seymour arrive together at Seymour's Sandwich Shop for lunch. While they eat, Seymour's postdoc Tai shows them his photographs of fly brains. Then they name the latest fly mutants after the foods that the flies' brain lesions remind them of: *egg roll, popcorn, spongecake, bubblegum, meringue, chocolate chip.*

# CHAPTER FIFTEEN

# The Lord's Masterpiece

*The Gods are here, too.*

—HERACLITUS

THE GENE *period* has a distinctive run of letters near the center, a run that repeats again and again: ACA GGT; ACA GGT; ACA GGT . . . This stretch of code produces a corresponding repeat of two amino acids in the *period* protein. There the repeat goes threonine-glycine; threonine-glycine; threonine-glycine; threonine-glycine; . . .

After Bambos Kyriacou left the Drosophila Arms and set up his own Fly Room at the University of Leicester in England, he and his students began collecting *melanogaster* from points all around the compass and examining this run of repeats. They collected flies from Bristol, England; Leiden, the Netherlands; Bordeaux and Saint-Tropez, France; Casablanca, Morocco; Andros, Greece; Rethimnon, Crete. Back in the laboratory, they sequenced *period* gene after gene. They found that not all of the strains had the same number of repeats: some had seventeen pairs, others twenty, others twenty-three. The farther from the equator, the longer the run; the colder the fly the longer the run.

Kyriacou knew that repetitions in the fly genome are common, just as they are in the human genome, where it is not unusual to find repeats of as many as 1,000 nucleotides. At one hypervariable site in the human genome, the sequence that repeats is only fourteen letters long, but the repeat length is 600 letters in some people, 1,200 in others, 2,200 in still others. People all over the planet carry so many different lengths of repeats like these that to find two individuals who have several different repeats of exactly the same length, one might have to look at the DNA of many hundreds of thousands of people. The molecular geneticist Alec

Jeffreys, who works in the same building as Kyriacou at the University of Leicester, has used these repeats to design the DNA fingerprint tests that are now produced as evidence and debated in courtrooms all over the world.

Kyriacou was interested in the repeated runs of ACA GGT in the center of *period* because he knew that the same repeat had been discovered in a clock gene of the fungus *Neurospora*. This stutter or stammer of DNA is the only sequence that the two clock genes have in common. Fly and fungus last shared a common ancestor at the start of the Cambrian period, the dawn of every great animal type from the invertebrates to the vertebrates. Fungus and fly have been evolving separately for 600 million years, which means that their two branches on the tree of life are separated by a total of 1 billion, 200 million years of independent evolution. Since they still carry this same run of repeated letters in the middle of their clock genes while everything else has evolved and changed, Kyriacou assumed that the repeats must be doing something fundamental and irreplaceable for the clock. This proved to be a good assumption; in fact, investigators have since found these same threonine-glycine repeats in the clockwork genes of yeast, mice, white laboratory rats, and naked mole rats.

To find out what the repeats might mean, one of Kyriacou's collaborators, Tony Tamburro of the University of Potenza in Italy, tried synthesizing the repetitive portion of the *period* protein in his laboratory. Synthesizers were now standard equipment for molecular biologists. Almost everyone had a DNA synthesizer, for instance, and the machine was almost as simple to use as a typewriter with four keys, A, C, T, and G. The biologist typed in the letters, and the DNA synthesizer cobbled together the gene. Many molecular biologists also had a peptide synthesizer. The biologist typed, and the machine cobbled together a protein.

So Kyriacou's collaborator, Tamburro, typed up the repeats in a *period* protein, and the repeats promptly coiled into a helix. Tamburro tried it over and over: always a helix. In a series of experiments, he found that any quantity of these repeats would make a helix and that three repeats is the minimum required to make one complete turn. Kyriacou realized that this might be why his variants tended to differ by three: seventeen in Casablanca, twenty in Bordeaux, twenty-three in Bristol.

By now Kyriacou was intrigued. He knew that with only one piece of only one gene, he could not hope to understand how the whole clock works. As the drosophilist Peter Lawrence of the MRC Laboratory in

Cambridge, England, writes in his book *The Making of a Fly,* "Attempting to study a process with only a subset of the genes involved carries a risk: like trying to understand how a car engine works from a few pieces—say, a piston, the main gasket and the bolts that attach it to the chassis." Nevertheless, Kyriacou began to think that the repeats might code for a sort of clock spring, and that flies far from the equator, by evolving one or two extra coils to their springs, might have become, in a sense, slightly overwound. He wondered why the colder flies would have added these coils.

Kyriacou found by further experiments that at warm temperatures, a clock with a short spring runs at very close to a period of twenty-four hours. That is, it keeps a beat that is very close to the period of the earth's rotation, without needing to be reset. A clock with more coils in its spring runs faster, so it needs to be reset every day. In warmer temperatures—an oasis in Egypt, an open-air market in Morocco—the short clock spring may be the favored model partly because it runs so close to twenty-four hours and a fly does not have to go through the physiological effort and trouble of resetting it. On the other hand, Kyriacou sees some evidence in laboratory tests that a clock with more springs is sturdier and keeps better time in the cold. In northern Europe, it may be worth the trouble of resetting the clock every day for the sake of having a robust clock that keeps good time in Bristol.

The short spring and the siesta may be the aboriginal forms of the clock, since fruit flies evolved in Africa. In Africa the greatest danger in the lives of flies is desiccation. They wake each day when it is light enough to see but still early enough for *Drosophila* the dew lover to get an easy drink. They always take a nap in the heat of the day. A nap is for them what faith is in Hebrew Scripture: "A defense from heat, and a cover from the sun at noon,/a preservation from stumbling, and a help from falling." The flies in the North have evolved their overwound spring since leaving home.

Of course, Darwin's process, which the evolutionist Richard Dawkins calls the Blind Watchmaker, is tinkering and experimenting with the watch even now. In fact, the watch seems to be a mechanism that fascinates—so to speak—the Watchmaker. Kyriacou and his colleagues have studied tens of thousands of descendants from several hundred families of flies. In one study they analyzed almost 40,000 individual *period* genes on 40,000 X chromosomes. They found that the mutation rate of the DNA that codes for the clock spring is tens of times higher than the average mutation rate of fly DNA. Apparently nature is busily selecting the best clocks all over the planet, selecting "daily and hourly," in Darwin's famous phrase from the *Origin*. In this way the Blind Watchmaker is maintaining exquisitely subtle differences throughout the clock genes of the flies in England compared to those of flies in Africa. The Blind Watchmaker is tuning and fine-tuning their watches every day, everywhere.

Our own species has not had much time to adapt to the new schedules of artificial light that we have imposed on ourselves all over the planet. The clock watchers in their Fly Rooms sometimes wonder if there may be lessons here for our own longevity. If we do not treat our clocks right, will they kill us or make us sick? This is not just a question of coping with jet lag or with changes between night shifts and day shifts. It is a question of following the dictates of our own personal rhythms from day to day. "If a man does not keep pace with his companions," says Thoreau in a quotation beloved of dreamers and nappers everywhere, "perhaps it is because he hears a different drummer. Let him step to the music which he hears, however measured or far away."

Benzer knows how dangerous it can be to interfere with his own eccentric circadian rhythms. Back in his gypsy days as an itinerant geneticist, when he worked at Oak Ridge National Laboratory in 1948, government rules required him to start work in the morning, like everyone else. He had two car accidents that year on his way to the lab. One fellow scientist who saw Benzer driving at that hour noticed the look in his eyes and determined to take a different route from then on.

Many people with normal clocks notice themselves fading out at about the same hour almost every afternoon, usually at three or four o'clock. If they are doing something interesting, they can bull through that dimming hour. But often clock watchers, when they are trying to read a dense paper in a field they do not know well, wonder drowsily whether midafternoon may be a window of time when most of us are *supposed* to be asleep, perhaps because our own species evolved in a hot climate . . . in Africa, along with the little fly.

IN 1988, after years of slogging, Rosbash and Young and their laboratories finally began to get their breakthroughs. First, they caught a glimpse of the *period* gene in action inside the clock. That discovery was made by Kathy Siwicki, a postdoc in Hall's lab. Siwicki harvested flies hour by hour and checked to see how much *period* protein they contained. She discovered that the protein's concentration inside a fly's head rises and falls with a daily rhythm. In 1990, a student in the Rosbash lab, Paul Hardin, discovered that *period* RNA also cycles.

In the rival Fly Room at Rockefeller, two of Michael Young's postdocs, Amita Sehgal and Jeff Price, after screening seven thousand lines of flies, found a new mutant that had something wrong with its sense of time: *timeless*. The *timeless* gene is on the fly's second chromosome. In 1995, another of Young's postdocs, Mike Myers, working twelve hours and more a day, managed to clone the gene, which meant that the Young lab could now sequence it and read every last letter. The *timeless* mutant is missing sixty-four letters of code. Once he had cloned the gene, Myers could also check the expression of the gene in the fly heads. He could do just what Siwicki had done: freeze fly heads hour by hour and see if *timeless* switches on and off too. "It's definitely the right gene," he told Young, "and it's oscillating."

So *period* makes a protein that oscillates; *timeless* makes a protein that oscillates; and when the clock watchers mixed the two proteins in a petri dish, they formed a strong bond. They mesh the way gears' teeth mesh in a clock.

The clock gene was rapidly becoming famous in molecular biology as a kind of exemplar of the work ahead. Everyone interested in genes was interested, at least casually, in the discovery of two working parts of a clock that would mesh. And the same year Sehgal and Price found *timeless*, Joseph S. Takahashi, a molecular biologist at Northwestern University, found a mouse with something wrong with its sense of time. He found it by Benzer's method: injecting his subjects with a mutagen and watching their descendants for the odd mouse that seemed to march to a different drummer. Virtually all of his mice were extremely punctual about waking up and running on their exercise wheels, but this mutant's clock had a period that was more than one hour longer than normal. Takahashi had thought he and his group would have to screen thousands of mice to find something like Konopka had found, but he struck gold with the twenty-fifth mutant mouse—another triumph for Konopka's Law. Takahashi and his team named their mutant *Clock* (Circadian Locomotor Output Cycles Kaput). In 1997, Takahashi cloned *Clock*. Three mouse genes, *mper1*, *mper2*, and *mper3* (mouse *period* 1, 2, and 3) turned out to contain long stretches of code that are very similar to the fly's. Soon after *Clock* was discovered in the mouse, the corresponding gene was found in the fly as well, a gene now called *dClock*. It maps to the sixty-sixth region on the fly's third chromosome.

For all of the clock watchers on the planet, whether they were working on the clock in flies or mice or ferns or moths, *Clock* was another cog in the machine, another clue to the working of the clock, which appears to be a mechanism that has been conserved (at least in its broad schematic outlines) since the beginning of time. On blackboards and whiteboards in three or four laboratories, researchers drew the molecular equivalent of a description of the innards of a universal grandfather clock. "We've cracked the case," Hall began telling the reporters who came around more and more often to knock on the old Dutch double doors and ring the little hotel bell of the Drosophila Arms and disturb his terriers. "Almost every article ever written about biological clocks starts out hypnotically, saying, 'Biological clocks are a total mystery!'" Hall would cry as they sat together, knees to knees, in his small office, surrounded by escaped clock mutants, Civil War paraphernalia, the scents of ether and molasses, and the agitated terriers. "They *still say that*! It's *no longer true*, OK? That is to say, it just simply *is not true* any more! We now have ways to think about how this clock is really ticking. And in fact, I would be willing to say, 'This is the *way* it is ticking!'"

In the current model, the genes *period* and *timeless* are both switched on inside the nucleus of a nerve cell in the brain. The finger of the angel, so to speak, is always inclined to hold that switch on. So the cell makes the *period* protein and the *timeless* protein. When concentrations of these proteins reach a critical threshold, they enter the nucleus and turn off the switch—turn off the very genes that made them. This cycle takes about twenty-four hours.

Once the switch is off, the *period* and *timeless* proteins gradually decay within the cell, and when their levels fall low enough, the gene flicks back on. The proteins that flick the gene back on are encoded by *Clock* and by another newfound gene called *cycle*.

Any living clock must have three basic working parts. There has to be an input pathway, so that dawn and dusk can reset the clock. Otherwise human clocks, for instance, which have a twenty-five-hour period in caves and windowless rooms, would cycle into and out of phase with the planet. In fact, that is just what happens to many blind people. The seat of the human clock is in the suprachiasmatic nucleus of the hypothalamus, just above the optic chiasm that figured in Sperry's studies of split-brain patients. Many blind people's clocks run out of phase with the world, rather like split-brain patients' when part of their thoughts run independently of the other half.

Today the rival clockwork laboratories are continuing to look for new cogs and study how they fit together. They are studying the input pathway; the clock mechanism itself; and the output pathway, by which the clock controls the rhythmic behavior of life, from our lying down to our rising up. For molecular biologists and drosophilists, *per* has fulfilled its promise at last. It is now a model of the way genes work together with the outside world and among themselves to keep a body alive. Philosophers used to ask whether we are born with or invent our sense of time. Now we know that we have clocks woven into every one of our cells. Philosophers also used to ask how we know that the sun will rise tomorrow. In a sense we have that answer built into us, too. The clock is a kind of orrery in the heart of every one of our cells, revolving to help us keep time with our world, a model of the cosmos inside our heads that cycles whether we are in or out of sight of the sun. The revolution of the stars and the seasons is written in the turns of our DNA.

The clock is also one of late-twentieth-century biology's best demonstrations of the family likeness of genetic mechanisms, even between animals as far apart on the tree of life as flies and fungi on the one side and mice and human beings on the other. Human beings carry a number of genes that are homologues, close cousins of *per*'s, and at least one of these human genes comes complete with the same threonine-glycine repeats whose evolution Kyriacou is studying. We also carry a *timeless* gene, and our *timeless* protein meshes with our *period* protein like two gears in a clock. Since the Cambrian period, all over the face of the earth, the clock genes have intermeshed with one another and also with the sunrises and sunsets overhead. We think of our set of genes as a letter in a closed envelope that we receive from the generation before us and pass on unopened to the next. But genes are not like that at all. The essence of these messages is to be read and reread, opened and closed continually, like open letters or scrolls beneath the silver finger of a ceremonial pointer, in a nonstop give-and-take that is the essence of life.

IN 1998, as these pieces of the clock fell into place at last, a molecular biologist in Switzerland (a country that appreciates fine watchmaking) praised the clockwork story in *Nature* for its heuristic value, a value that makes it an instant classic in the annals of molecular biology. The now-celebrated gene *period* also stands as a corrective to the hype that surrounds the gene and the Human Genome Project by showing that single genes are usually not the answer, and that even a consortium of laboratories will often have to struggle a long time to figure out how a gene complex works.

More often than not, this is the way these detective stories will run in the future. It is the way the study of the obesity gene complex is running, for instance. There the discovery of one gene (in a lab across the hall from Michael Young's Fly Room at Rockefeller) was followed almost immediately by the discovery of many more that intermesh with it.

The gene that causes Huntington's disease is another case that parallels the story of *period*. After a struggle as long, and often as agonized, as the struggle with *period,* it has now been cloned and sequenced. It, too, has a conspicuous repeat embedded in it. Most of the repeats in the human genome seem to be as harmless as the whorls on fingerprints. But certain unstable repeats of three nucleotides can produce fragile-X syndrome and myotonic dystrophy, both of which cause mental retardation. Certain longer repeated units can lead to cancer and diabetes. And in the mid-1990s—a dozen years after the gene was mapped—molecular biologists began to suspect that it is the repeat in the Huntington's mutation that does the damage.

The repetition is a single codon: CAG, CAG, CAG. Human beings carry many different lengths of this repeat, from fewer than twenty to more than forty. CAG is genetic code for the amino acid glutamine. A gene with extra repeats makes a protein with a long string of extra glutamine molecules one after the other. A normal *Huntington* gene makes a protein with a run of nineteen to twenty-two of these repeats. A mutant protein has from about forty to hundreds of repeats.

Like the *period* gene's protein, the *Huntington* gene's protein works with a second protein that acts as its partner. Its partner is a protein called glyceraldehyde-3 phosphate dehydrogenase, which is an enzyme that does many kinds of housekeeping work in the cell. When the *Huntington* gene's protein carries many extra repeats, it binds more tightly than normal to its partner, like two gears jamming, in what eventually becomes a fatal embrace. In other human genes, these same CAG repeats can cause other problems, a whole *Merck Manual* of nervous disorders, including various kinds of ataxia and muscular atrophy.

In this way, by following the molecules, biologists are beginning to trace some of the first links from human molecules to human behavior, beginning with aberrant behavior, as in Benzer's Fly Room. But even at the century's close, molecular geneticists still knew so little about the

*Huntington* complex that Nancy Wexler, a biologist who had helped begin the search for the gene and whose mother had died of the disease, refused to take a test to find out how long her own number of repeats might be. No one could do anything to help her if she had a long number of repeats, and she decided that she did not want to know.

The dream of this research is the cure of diseases like Huntington's. Those cures may come in time, but the lesson of *period* counsels patience. In this lesson there is also some consolation. If the dream is still some distance away, then so is the nightmare. The nightmare is the development of the kind of genetic engineering project that would diminish whole generations' sense of free will. Everyone who works in the field hears questions like "Have you found the free-will gene yet?" It is a joke that contains the essence of the late-twentieth-century horror that Benzer's science has opened. And the assumption behind the nightmare is that someone could engineer or reengineer that gene as easily as a genetic engineer can now change the clockwork gene and transform a fly's sense of time.

However, the actual clockwork of the fly is turning out to be so complex—and more pieces of the clock are still to be discovered—that it looks less like an invitation to human intervention and more like a cautionary tale or object lesson for anyone who might try, in the twenty-first century, to improve on nature's four-billion-year-old designs. Molecular biologists have now been mutating and examining thousands of generations of flies, and they have not made a better clock or a better fly yet.

In 1962, the year Benzer began to think about his study of genes and behavior, the novelist Anthony Burgess published *A Clockwork Orange*. Burgess later explained the title: "I mean it to stand for the application of a mechanistic morality to a living organism oozing with juice and sweetness." Even then, when most scientists thought in terms of nurture rather than nature, Burgess was worried that a science of behavior and mechanistic morality might rob us of free will. *A Clockwork Orange* is the story of a young convict forcibly conditioned by the methods of behaviorists to abjure violence. The novel explores the nightmare of total control of behavior from outside, the way science fiction novels today explore the nightmare of total control from the inside. Burgess's prison chaplain doubts that the reprogramming of the novel's antihero will constitute a real improvement: "The question is whether such a technique can really make a man good. Goodness comes from within, 6655321. Goodness is something chosen. When a man cannot choose he ceases to be a man."

And again: "What does God want? Does God want goodness or the choice of goodness? Is a man who chooses the bad perhaps in some way better than a man who has the good imposed upon him? Deep and hard questions, little 6655321."

ONE OF THE BEAUTIES of the clock for molecular biologists has always been its neutrality. A clock is not a loaded subject—except as a symbol for all the rest of the living machinery that the fly clock is now teaching scientists to study, much the way the deconstruction of clocks and flies has taught generations of young boys to become scientists.

Of course, the other mutants that came out of Benzer's Fly Room are not emotionally neutral; and in the 1990s, as these studies advanced too, gaining by monthly advances in the powers of molecular biology, the Benzer school's work at last began to emerge into the light and heat of the world's attention.

Jeff Hall and his technicians have a Fly Room they call Paradise, because they keep it at a constant Bahamian temperature. In Paradise, *fruitless* flies in rows of glass vials go round and round, chaining from dawn till dusk. Hall is now working with a whole consortium of other Fly Rooms on the genetic dissection of *fruitless*. The mutation is turning out to be embedded in a gene complex at least as intricate as the orrery of *period* and much more sensitive politically.

Hall had gotten an early hint of the complexity of the instincts involved here when he made his gynandromorphs in Benzer's Fly Room. If a gynandromorph has a female body and a male head, the head says, "Court," while the body says, "Be courted." The gynandromorph with the female body extends a wing and sings to females but sings a sort of gibberish song—unless there are patches of maleness in the thorax.

Hall and his team have now generated a long line of *fruitless* mutants. In essence, what they have done is to create a library of artificial *fruitless* alleles. They have fed innumerable strains of *fruitless* EMS and bombarded *fruitless* with gamma rays, all to make more mutants. Each group of *fruitless* mutants lives together in a vial for a few days before they start to chain. A technician in Hall's lab monitors how long they take to chain and how much they chain when they do. Some vials are more *fruitless* than others. Depending on the damage in the genes, just a few flies in a vial may make just a few circles now and then. They break up and then come back together; they take little breathing spaces. They extend their wings, but not fully, making scissoring motions, almost as if they cannot get their wings out far enough or up high enough. Other vials seem to go crazy: with one wing out and shaking, the mutants dance in circles all day long in what looks to human eyes like a frenzy, waving their wing tambourines.

In 1989, *fruitless* was mapped by a postdoc in Hall's laboratory. The mutation lies on the fly's third chromosome. The original mutation was made by zapping flies with X rays. Now Hall could tell exactly what those X rays had done to the chromosome. The X rays had damaged it in two loci that lie very close together. A piece of DNA had popped out and then gotten stitched back into its chromosome backward. The X rays had caused an inversion, and this inversion has been passed down in every *fruitless* fly. In effect, *fruitless* is two mutations close together, one at each end of the inversion. Some of *fruitless*'s behavior maps to the break point at one end of the inversion: the *fruitless* male's failure to copulate and his habit of chaining with other males. Another piece of *fruitless* behavior maps to the other end: the male's ability to stimulate courtship in other males.

In the 1980s and 1990s, the pursuit of *fruitless* led Hall and others into a deep study of the origins of the differences between the sexes. As with the study of *period*, the findings turned out to have great generality. Why does one fertilized egg grow into a male nervous system and another into a female nervous system? What are the molecular events that determine sex? It was Morgan's Raider Alfred Sturtevant who discovered the first sexual determination mutant, a gene he called *transformer*. When a female fly inherits two copies of *transformer*, her chromosomes remain female, and she still grows big, like a female, but otherwise she looks and acts like a male, with a male's black belly and a male's gift of song. Mutations like *transformer* have since been discovered in every imaginable kind of organism, including human beings.

What *period* is to the sense of time, *fruitless* is to sex: another demonstration of Konopka's Law. It is a point of entry that is helping to lead molecular biologists inside what is now one of the best-studied stories of gene interaction in nature. Molecular studies have uncovered a whole cascade of genes that decides the sex of the fly. The point of origin of the cascade is a gene called *Sex-lethal* (discovered by another Raider, Hermann Muller). If a fly inherits two X chromosomes, then *Sex-lethal* is turned on. *Sex-lethal* turns on *transformer*; *transformer* and *transformer-2* turn on *doublesex*; *doublesex* turns on a cascade of genes that give a fly a vagina, her feminine perfumes, and all the other sexual equipment of a fit female fruit fly.

On the other hand, if a fly inherits only one X chromosome, the cascade flows on a different course to *doublesex,* and *doublesex* then switches on a different cascade of genes. This cascade gives a fly a penis, masculine perfumes, and the celebrated sex combs that decorate the entrance to Hall's laboratory, the Drosophila Arms.

Now, *fruitless* turns out to be one of the key genetic switches in this long cascade. The gene is in the category of tremendous trifles. As the apostle James observes, there are tiny objects in this world that are able "to bridle the whole body." "Behold, we put bits in the horses' mouths, that they may obey us, and we turn about their whole body." The gene *fruitless* is such a bit. Some of this gene's products are expressed only in the central nervous system, and only in about 500 out of the 100,000 neurons in the central nervous system. Of those 500, most cluster in groups of 10 to 30 cells, and they map to the sections of the nervous system that are required for carrying out the dance steps of courtship.

Sex in our kind is determined by cascades of genes too. A gene on the short arm of the Y chromosome starts the building of the testes and makes a human embryo male. This gene was cloned in 1990 and named *sry* (for "sex-determining region, Y chromosome"). Mutations in *sry* can cause babies with two X chromosomes to develop as boys and babies with an X and a Y to develop as girls. The more molecular geneticists study these mutations, the better they will understand exactly what happens to make a man look at a woman or a woman look at a man and see the lineaments of gratified desire. These lineaments begin in lines of genetic code; they become elaborated in rich and complicated ways as the embryo grows, responding to the embryo's inner and outer environments, Pascal's two infinites.

In *What Is Life?* Schrödinger imagined that the physical structure of the gene would turn out to be the Lord's masterpiece. But the gene itself is only the beginning. The true masterpiece is in the procession of interconnections of small wheels within bigger wheels within bigger wheels that make a life, a procession that merely begins with the lines of code in the gene. The more biologists study these wheels within wheels, the more complexity they find. Years ago the evolutionary biologist Julian Huxley warned of the danger of anthropomorphizing animals and the equal and opposite danger of mechanomorphizing them. We lose sight of the animal either way, he said: by imagining it is just like us or by imagining it is just like a machine. Now we can see the truth of this observation at the level of the clockwork mechanism itself, with *period,* and equally, with *fruitless.* At each turn, inside the growing embryo, a gene interacts with the genes around it, and in the adult animal each complex of genes interacts with many others and with the surrounding world in ways so complicated that we are only just beginning to explore them, even in a fly.

# CHAPTER SIXTEEN

## Pavlov's Hat

*If knowledge isn't self-knowledge it isn't doing much, mate.*

—TOM STOPPARD,
*Arcadia*

WHEN CHIP QUINN left Benzer's Fly Room, he took the memory project with him. In a Fly Room of his own at Princeton University, he went
on poisoning flies and running their children through his Teaching
Machine. In this way he found more slow learners to keep company with
the original *dunce,* including *amnesiac, smellblind, turnip,* and *rutabaga.*
But Quinn was groping. Memory was a black box, and he did not know
how to get inside it. As one of Benzer's student's students observed afterward, *dunce, turnip,* and *cabbage* all showed "essentially the same overt
phenotype: stupidity." Quinn could not distinguish whatever was wrong
with *dunce* from whatever was wrong with *rutabaga.*

One day a grad student of Quinn's, Ronald Booker, fixed a fly to the
end of a wooden stick with a dab of wax. He tied a slipknot in a fine wire
to make a noose and looped the wire over one of the fly's legs. He
cinched it tight by pulling on it with a forceps and snipped the wire so
that one end dangled loose. Then he suspended the little fly over a pool
of salt water so that the wire just brushed the surface. The fly waved and
kicked its leg, and the wire lifted out of the water and dipped back in.
Whenever the wire touched the water, the fly got an electric shock. Normally, Quinn's wild-type flies were uncertain learners, as Benzer's neme-
sis Jerry Hirsch had pointed out. But hanging over the electrified pool,
better than nine out of ten wild-type flies learned to keep that leg up.

Then Booker tried pinching off the fly's head with a hot forceps or
cutting it off with a razor blade while the fly hung over the water. Not

only did the fly survive decapitation, it learned to keep the wire out of the water even better than a fly with a head. Flies have nerves outside the brain, just as human beings do, and apparently those nerves were learning the lesson. Quinn and Booker could only assume that the flies learned better without their heads because they were less distracted.

However, *dunce, turnip,* and *rutabaga* flunked this leg-lifting test with or without their heads, just as they flunked all of Quinn's other learning tests. Something was different in both their genes and their nervous systems. But to find out exactly what, Quinn wrote in 1981, "we will need more sophisticated tools than hot forceps and a razor blade."

That year, Quinn took on a new postdoc, Tim Tully, who was probably as passionate about the study of memory as Quinn had ever been. When Tully was a boy, a day of childhood memories had been knocked out of him in a Christmas accident. Hitting a tree with his sled had produced a kind of hollow of amnesia; he had lost all memory of the hours preceding the accident. (A quick dunk in ice water will do the same thing to a fruit fly.) When he was in college, studying with Jerry Hirsch, Tully decided that finding the secret of memory was the only thing he wanted to do in his working life. And after studying the literature he decided that Benzer's method of genetic dissection was a more promising tool than Hirsch's method of breeding and crossbreeding flies. He told Hirsch that Benzer was discovering molecules; Hirsch was not. Tully often says that Hirsch regarded him as a reincarnation of Judas for going over to Benzer. "And in a way perhaps I was," Tully says. "Although I never kissed him on the cheek."

Tully joined Quinn at Princeton and took stock of Quinn's operation. He was just as blunt with Quinn as he had been with Hirsch. In his view, Quinn's problem was the inefficiency of his screen. Quinn's headless fly had proven that flies are capable of much better performances than they ever achieved in Quinn's Teaching Machine.

Tully has a knack for gizmos and gadgets. He designed a new Teaching Machine. He piped in the odors silently and smoothly on soft breezes so that all the flies would get the same doses of his perfumes. He lined the shock tubes in such a way that every fly would feel every jolt. He designed each part of the machine to run quietly and efficiently so that the flies would not be distracted from their lessons and so that virtually every one would get virtually the same lesson every time. He used a little plastic elevator to lift the flies gently from one stage of the experiment to the next. Designing and building this machine took Tully four years.

Human eyes can see dim red light, but fly eyes cannot. So when Tully was ready to run his new Teaching Machine, he turned out the lights and watched the flies by the glow of a low-watt darkroom safelight. In the dark and quiet of the new machine, fruit flies acted much calmer than they ever had in Quinn's, and Tully was delighted to find that better than nine out of ten of them now learned their lessons. Tully had gotten rid of the static—which is what Quinn and Booker had done much more crudely with the hot forceps and the razor blade.

Now, because the flies in the new Teaching Machine were not distracted from their lessons, Tully's experiments were not blurred by the random noise of good learners that failed to learn. And now Tully could discern varieties of stupidity, a wide spectrum of troubles in Quinn's mutants. He discovered that *rutabaga,* for instance, is able to learn; it just forgets fast. In other words, *rutabaga* is a memory mutant but not a learning mutant. Other mutants have trouble learning, but they hold on to what they do learn. They are learning mutants but not memory mutants.

During those long early years of R and D, Tully had the same public relations problems that had always plagued Benzer's school. Most biologists found it hard to take a fly in a Teaching Machine seriously, and Tully had trouble getting any funding for his project. When James Watson first got interested in Tully's work, Watson's advisers urged him to stay away. But Watson installed Tully in the laboratory next to his office at Cold Spring Harbor, and slowly Tully began dissecting the act of memory in his flies.

For years, Tully had been reading the papers of the neurobiologist Eric Kandel, who is now at Columbia University. Kandel was working on learning and memory in a giant snail, *Aplysia.* When Kandel tickled the siphon of the snail, it flinched and contracted its gill. But when Kandel tickled the snail ten or fifteen times, it flinched less and less. The snail was learning to ignore him. If he smacked the poor snail on the head, it suddenly became much more responsive again.

Kandel and his colleagues managed to trace the circuit of neurons that governs these simple reflexes. The team stuck a microelectrode into a single neuron and recorded the characteristic pattern of bursts of electricity, like a series of dots and dashes in Morse code, that made the gill flinch. In one experiment, Kandel sent precisely that same series of spikes into a motor neuron. The gill flinched. Kandel was learning the snail nerve's language, the way King Solomon in the legend learned the languages of the beasts, fowl, and fishes.

From there Kandel played with some of the molecules that compose the message; this was the work that fascinated Tully. One nerve sends another nerve a message across the synapse between them. The message takes the form of a packet of molecules—or a burst of packets. When the snail gets knocked on the head, a message shoots down a nerve from its head and activates an enzyme called adenylate cyclase in the gill nerve. This enzyme helps the gill nerve make a second compound called cyclic adenosine monophosphate, or cAMP. The cAMP then cranks up the nerve cell's sensitivity so that the next time the siphon is tickled, that cell will release more packets across the synapse. In this way, the memory of the blow to the head has left a trace in the snail: the snail will flinch harder the next time it is tickled in the siphon because it has just been struck. It is a simple case of learning and memory, reduced to a few molecules.

At Caltech, Benzer and his student Duncan Byers were following Kandel's work too. Since cAMP matters so much in the learning and memory of the snail, they decided to take a second look at their fly mutant *dunce*. They were thrilled to discover that *dunce* makes a crippled form of the enzyme cAMP phosphodiesterase. Without that enzyme, *dunce* has problems metabolizing cAMP. So at this level, at least, the language of the fly and the snail seemed to be the same. Benzer and his students wondered if the enzyme cAMP might be part of a universal mechanism of learning and memory in the living world.

Meanwhile, to study the formation of long-term memory in his new and improved Teaching Machine, Tully gave a set of wild-type flies ten lessons one after the other, without a break between them, as if the flies were cramming for a final exam. He gave another set of wild-type flies ten lessons with rest periods in between, as if they were preparing for their exam by studying every other day. The class of crammers forgot everything fast; they made only short-term memories. But the class that took breaks still remembered the lessons after a week.

During the first half of the 1990s, Tully and a growing team of colleagues, postdocs, and graduate students defined step after step in the making of these lasting memories at the molecular level. They did the work by using the old memory mutants in ingenious new combinations in Tully's Teaching Machine, along with some brand-new mutants that he and his lab discovered along the way, including *latheo*, named after Lethe, the river in Hades that caused all the souls in the underworld to forget their past, and *linotte*, named after the French expression *tête de linotte*, for which the English equivalent is "birdbrain."

Kandel and his laboratory had discovered that cAMP turns on certain key genes involved in making permanent memories—but which genes, they did not know. Their work suggested that a certain protein—cAMP-responsive element-binding protein—might be involved in the cascade. The gene that makes this protein is known as *creb*. In 1994, one of Tully's chief partners in this work, Jerry Yin, cloned one of the fly's *creb* genes. The *creb* gene can make a protein in one of two alternate forms, and these two forms have opposite functions in the brain. One form activates certain genes; the other deactivates those same genes. In other words, one form is an on switch and the other form is an off switch.

When they found that switch, Yin, Tully, Quinn, and a few other collaborators began one of the most remarkable genetic engineering projects in the study of genes and behavior. They mixed a DNA transformation cocktail containing a *creb* gene: the on switch. They injected this cocktail into the eggs of wild-type flies. When they had made and bred these transgenic flies, they loaded a class of them into the Teaching Machine.

Each fly in that classroom carried an extra on switch. What is more, Tully and his team had mixed the DNA transformation cocktail in such a way that the extra on switch would flip on only when the fly was warm. This is one of the routine tricks of genetic engineering that are now possible in Fly Rooms; the gene involved is called a heat shock promoter, because heat will shock the gene into promoting the action of the other genes in the DNA transformation cocktail.

In a cool room these flies learned normally, as expected. Now came the test. Tully took a bottle of the flies and dunked it in warm water. Inside the brains of those flies, the extra switch flicked on. Three hours later, Tully gave the flies a single lesson. The flies remembered that lesson for a week. And as one drosophilist wrote in a commentary on Tully's work, "For flies in the wild, one week is long enough to arrive at midlife crisis." It was as if a boy of twelve were told a phone number once and he still remembered the number at the age of forty. With the extra on switch, the flies had the equivalent of a photographic memory.

Tully and his lab had proven that they had found at least a piece of the machinery of memory. In the same stroke, they had changed memory in the direction that every thinking generation has dreamed of.

FOR AN ARROWSMITH, this discovery, building as it did on a century of cumulative scientific effort, would have brought blurred nights of almost religious awe. For Watson, who strains every nerve every day to support the Cold Spring Harbor Laboratory, the significance of the moment was more mundane. According to laboratory legend, when Tully first told Watson, Watson screamed out, "We're going to be rich!"

Watson had come to agree with Benzer that the path through the gene would lead to a greater understanding of the mind and brain. But unlike Benzer, Watson thought immediately of the control of illness, the enhancement of the mind, and the commercial and political possibilities of anything that pointed in those directions; which is why he had decided to gamble on Tully. Soon after Tully's discovery he was talking it up to Wall Street investors, with one of his self-mocking smiles: "*Insect brains.*"

"Well, you see, genetics in a *sense* restricts our freedom," he says, sitting in the vast suite of offices that he inhabits as director of the Cold Spring Harbor Laboratory. "I mean, we're born with a lot of things, which means that we can't do *everything*. And some people have more freedom than others. It's not an—" A pause. "It's, um, slightly uncomfortable," he adds in a confiding tone. "It's *really* uncomfortable if you don't have them, and if you *do* have them, wondering whether you *deserve* them. . . ."

Watson has an acute sense of himself and what he represents. His Nobel Prize certificate hangs on the wall behind his desk. The bell tower that tolls the hours outside one of his office windows has a staircase shaped like a double helix. To Watson it seems an omnipresent fact of personal experience that genes give some of us more freedom than others. "Well, the freedom not to be sick," he says, "or the freedom to have your brain work right. As distinct from most people," he adds, "whose brains *don't* work right."


BECAUSE OF THE family likeness of genes in the tree of life, no one at Cold Spring Harbor was amazed to discover that mammals have a *creb* gene very much like the fly's. In another laboratory at Cold Spring Harbor, a molecular biologist named Alcino Silva tried the same sorts of experiments with mice that Tully had practiced upon his flies. By now mice had reached the point where flies were in the 1960s: they had become powerful tools of genetics. Anyone with a computer in the mid-1990s had access to maps with 7,377 markers on twenty pairs of mouse chromosomes. If Silva found an interesting quirk in the memories of his mice, he could quickly track it down to a gene on a chromosome. Then he could clone that gene, insert it into a mouse egg, and see if it transformed the behavior of the mouse, just the way Tully and his crew had done with flies. The mouse genome is about the same size as the human genome, and for virtually every gene in us there is a corresponding gene in the mouse. It is true that the genes do not always do the same things, but gene for gene we are clearly quite closely related. Even our chromosomes are the same: if you chop mouse chromosomes into about two hundred pieces and put them together rearranged, you can make human chromosomes.

Not only can mouse workers take a single human gene and put it into a mouse and see what it does, they can even take a big fragment of a human chromosome from a human being—a fragment bearing hundreds or even thousands of genes—insert it into a mouse egg, and see what it does. And even before the first cloning of mice in 1998, hundreds of inbred mouse strains were so nearly identical that they provided geneticists with virtual armies of identical twins.

The *creb* gene in a mouse is an on switch and an off switch virtually identical to the fly's. So Silva tested a strain of mice with a mutant *creb* gene. In one experiment, he tried putting a mutant mouse in a round, steep-sided pool called a Morris Water Maze. When the mouse found itself in a Morris Water Maze, it tried to save itself. The only escape was a small island that was submerged just under the surface of the dark, opaque, deep-dyed liquid somewhere in the pool. After a few swims a normal mouse could learn to find the island. But the mouse with the mutant *creb* switch searched for the island at random all through the pool again and again. Every time, it made land purely at random. Without *creb* it could never remember where it had found the island the time before. It may even have forgotten there was an island to be found.

At Columbia, Kandel and his group were testing *creb* mice too. They used a white disk with holes around the rim, one of which was an escape tunnel that led back to the mouse's cage. Under the pressure of bright hot lights and a loud alarm-clock buzz, each mutant mouse dashed around the disk looking for the one hole that it had found before, the hole that would lead home. A mouse with a ruined *creb* gene would race from hole to hole, leaving signs of anxiety everywhere it went. It would sniff around between the holes, run back to the middle of the disk, scratch its head. When test time was up, a *creb* mutant would still be scurrying around looking for that one hole that led home.

So *creb* seems to be a universal sort of toggle switch that an animal throws when it wants to remember something for a long time. Apparently these switches have been preserved for many hundreds of millions of years. No one yet knows how many other switches there are that allow a human being to remember events in uncanny detail for uncanny amounts of time. Jeff Hall has this gift, though he is not always happy to have it. With each year he looks more burdened by memories, lost causes, and unrighted wrongs. His father was the same way: a star reporter for the Associated Press who never took a note. Hall is a prodigy on the Gettysburg battlefield, which he haunts in his off-hours in his stiff blue Union cap. Whenever anyone asks him a question, he answers volubly, voluminously, and with something so much like gratitude in his voice that he sometimes draws crowds.

Hall once wrote an angry letter to Benzer citing Benzer's lack of generalship, his defection from the field of behavior, and many other grievances. "I remain offended," Hall wrote, and added, "unfortunately, I never forget a thing."


THE CREATION of a fly with a photographic memory got so much press attention that one day in April 1995 Watson and Tully actually began to wonder if the work might be putting their lives in danger. Local, national, and international television and newspaper reporters were trooping through the Fly Room to see the flies perform in Tully's Teaching Machine. And in the middle of the media storm, the Unabomber, who had just mailed his sixteenth bomb, mailed a letter to the *New York Times*, a letter written in the royal or revolutionary "we." An excerpt of the letter ran in large type across the top of the front page. The letter explained the bomber's hatred of genetic engineers, among other specialists:

> We have nothing against universities or scholars as such. All the university people whom we have attacked have been specialists in *technical fields*. (We consider certain areas of applied psychology, such as behavior modification, to be technical fields.) We would not want anyone to think that we have any desire to hurt professors who study archaeology, history, literature or harmless stuff like that.

Watson dropped his copy of the *Times* on Tully's desk with the headline facing Tully. "So that mad bomber particularly doesn't like geneticists," Watson said, looking very pale and tired. He and Tully stared at each other. In their adjoining offices they felt like sitting ducks. "I think we've got to ask that all our packages go through an X-ray machine," said Watson.

"Oh, man," Tully said, "I'm history. It's just a matter of time before I lose my fingers."

But that spring Tully exulted in the success of his experiment. "I've always known exactly what I wanted to do," he said. "I didn't know how long it would take and how resistant people could be when they have you pigeonholed." Pigeonholed as hopeless—Watson has files full of letters advising him that it would be impossible to learn anything important about the human mind by poisoning the brain of a fly. "That has been frustrating. The whole miracle was Watson. It was an absolute angel that swooped down and said, you come to Cold Spring Harbor."

And that spring Watson exulted too. "If I were a sort of graduate student now, I'd probably gravitate toward something like Tully's lab," he said. Most young molecular biologists are working on the genes that control the growth of embryos. "I don't regard that as—" Watson snickered sarcastically. "You know, I don't want to work on the twenty-third homeobox." Let someone else nail down the details of the gene cascades that make an egg grow into a fly, a zebra fish, a mouse, or a human fetus. "Exactly how it grows—we're a long ways from totally describing that," Watson said. But the process is no longer essentially mysterious. "You could say that sequential gene action is the basis of it all. You've got these cascades that produce the eye . . ." And he rattled off some of the fantastic gene-to-gene cascades that shape the eyes and the limbs and differentiate the head from the tail. This is the subject that Brenner and Benzer helped to open up when they turned away from behavior, and it is the field that *fruitless* is now helping to illuminate.

"I mean, Jesus, embryology's solved!" cried Watson with dismissive hyperbole. "But I don't know how memory works in the slightest. You know, no one can give *any model* of the human brain. And that's why it just seems to me"—he snickered—"a bit of a challenge."

A finding like Tully's explains why so many of the first molecular biologists turned from the gene to the brain in the first place, Watson added. Molecular biology had been getting routine. "Whereas the brain is slightly different. It just seems much—more fun."

WITH THE CLONING and rewriting of the fly's *creb* gene, it was clear that Tully was inside his problem. He was exploring from the inside the kinds of behavior that Pavlov, for instance, had studied from the outside when he conditioned dogs to drool or flinch at the sound of a bell. Just as Pavlov's work became a cornerstone of behaviorism, which is the study of behavior from the outside, Tully's work will be a cornerstone of the new study of behavior from the inside. It is now quite clear from studies in both these fields that all animals, even flies and snails, can learn to associate two events that come at them one after the other: a bell and a morsel of food; a smell and a shock. With flies, dogs, or human beings, when an experimenter slowly separates the two events, learning slowly falls away. In each species there seems to be some maximal time span after which there is no further associative change. Tully believes that this limit reflects a cellular property of neurons that can now be studied at the molecular level and will eventually be found to function essentially the same way across all life. What Pavlov explored in his dogs and Tully and his colleagues are now exploring in flies and mice, the power to learn and remember, is very old. Since we have the on-off switch, and so do flies and worms, it must be ancient, like sex and the clock. This, too, is a molecular invention that is older than the Cambrian explosion; and it is hard to imagine a more interesting molecular invention for us to begin to understand.

Tully once made a pilgrimage to Pavlov's old institute in St. Petersburg and asked to see a list of the names of Pavlov's dogs. The only dog's name that he had ever been able to find in the literature was Birka, which means "Whitey." No one at Pavlov's former institute knew the dogs' names either. Then Tully went to Pavlov's apartment, which has been preserved just as he left it the day he died. A curator gave Tully a tour of the place. At the end of the tour, they sat down for stale coffee and biscuits, and the curator pulled out an old photo album filled with the names and mug shots of Pavlov's dogs.

Now Tully plans to honor those names. In his view, behavior sits at the top of a biological pyramid. At the bottom of the pyramid there are genes in a cell. Then there are cells interacting to make a living system: say, a walking, talking, learned, arguing organism. Learning and remembering are among the most exquisitely complicated feats that a walking, talking organism can accomplish: the absolute top of the pyramid of organization. So there are bound to be hundreds of genes involved in learning and remembering, and Tully plans to find them all, one by one, and name them after Pavlov's dogs.

When he and his lab engineered the fly with the photographic memory, Tully felt an amazing rush of vindication and exhilaration. They had made a beginning. He knew that much of his crew's ability to do such

*"It really is a pilgrimage." Tim Tully in Pavlov's old apartment in St. Petersburg, wearing Pavlov's hat. Benzer in Darwin's old house, south of London, writing at Darwin's desk.*

interesting work derived from their position in time, high in Occam's Castle. He thanked his stars that he had arrived on the scene just as it had become possible to grasp some of the first solid links between genes and behavior and to fulfill the mission he had been pursuing since he had been a student in college. "This is the point in time when we get to talk about these things," Tully says. "This is the defining moment. Things finally have come to a head." In one corner of his office at Cold Spring Harbor, he keeps a photograph of himself in Pavlov's old apartment, wearing Pavlov's hat. A few weeks after his eureka he crossed his office and peered at the photograph. "It really is a pilgrimage," he said. "You can see it on my face."

In the same spirit, Benzer once visited Darwin's house in the village of Down, not far outside London. He explored the place with the impudent curiosity he brings everywhere. His daughter Barbie was with him. She is used to her father's ways; she remembers following him as he strolled past "No Trespassing" signs in O'Hare International Airport because he wanted to see the control tower ("Where everyone else gets ushered out, he gets a tour."). At the Institut Pasteur, Benzer once outraged the institute's French administrators by asking if he and his family could have Pasteur's old apartment. (Absolutely not!)

Darwin's study was roped off. Benzer admired Darwin's chair, equipped with wheels so that Darwin could roll around among his files without getting up, and a blanket to keep him warm. "Barbie so impressed the guard that I got to sit in the chair," Benzer says.

"*No one* ever gets to sit in that chair," Barbie says.

The guard lifted the velvet rope, and Benzer climbed up into the chair. It was so high that his feet hardly touched the floor. "That was a great thrill for me," says Benzer. "They didn't have his slippers around, unfortunately, so I couldn't put on Darwin's shoes."

# CHAPTER SEVENTEEN

# Rough Mountain

*Felicity is a continual progress of the desire, from one object to another, the attaining of the former, being still but the way to the latter.*

—THOMAS HOBBES,
*Leviathan*

BENZER LEANS BACK against the wall of his laboratory's meeting room and sips his first cup of laboratory tea. A menu cover on the wall, an old souvenir from an extinct delicatessen, still says SEYMOUR'S SANDWICH SHOP. He wears a rumpled white lab coat over a button-down shirt, a woolen sweater, a loosened tie, the working clothes of the old school since Arrowsmith. Horn-rimmed glasses hang from a white cord around his neck.

Having worked until dawn in his sanctum, as usual, he is settling back to work at one in the afternoon, also as usual, nursing his tea while the young biologists around him chew the last of their lunches. Successful laboratories are like Renaissance studios: generations of students pass through them to learn from a master. Although Benzer retired from classroom teaching when he reached the age of seventy, young Ph.D.s still apply from around the world to work with him in the lab. His current crew comes from China, France, Germany, India, Japan, Korea, and Pakistan. At the moment Benzer himself is the only American in this group.

He has been a legend in biology since before his latest students were born—ever since *rII*. Postdocs study his face now the way Arrowsmith studied Max Gottlieb's. They give him their papers to read, and then they bring back his black and spidery marginalia for him to decipher ("almost unreadable, even to me"). They learn how to calibrate his praise, and they know the look with which he dismisses anything that is not solid, not well grounded, not at the root of things, not part of the foundation: a skeptical, fine look, a very slight rolling of the eyes and a slight smile.

Every week or so he invites seminar speakers—friends, famous and obscure—to help his lab follow the explosions of molecular biology. Every Friday afternoon he also herds the postdocs out of the lab and across the campus to the Red Door Cafe for coffee and conversation. Sometimes the postdocs exchange looks before they go. They are a new, driven generation of molecular biologists. For them Benzer's custom has the odor of the Old World, of Paris and Cambridge, of a more languid and leisurely civilization. One Friday not long ago, Benzer asked them if they had ever read *Arrowsmith*. The crew looked blank. A few of them knew Aerosmith, the rock group. Only one of them had read the book, and he pulled from his memory the name Leonora. Benzer cried sharply, with real pain in his voice after sixty years, "*Leora!*"

Sometimes lately he remembers what Salvador Luria told him the night they met—the night he saw Delbrück's picture, the night he left physics. Luria said something that seemed to him to be extremely important, although Benzer could not follow it at the time. "He said, 'Everyone keeps going down, down, down, trying to be more reductionist, trying to see finer and finer, to find the basis of structure and function.' And he said, 'I think it's time to start going up again—going in the opposite direction.'

"So I was interested in that," Benzer says now. "But of course it took a long time before I—it takes a long time of going down before you start looking to go up again. Down is a much easier way to go."

Now he is using the latest gleaming tools of molecular biology to work up from the gene to the trait and to work down from the trait to the gene. He is pursuing some of the mutants that he found in the first years of the Fly Room, and the excitement of the work is keeping him in the lab all night, night after night. Some of these mutants he has been trying to understand for thirty years; and now, one by one, he can. He follows Changsoo Kim, the Korean postdoc, out of the Sandwich Shop and into his laboratory darkroom, sipping his tea from a paper cup. "Oh, my God!" he cries, looking at Chang's microphotographs. "What's wrong with these guys? What are all those spots?"

Through the microscope, Chang has been making a series of micro-portraits of *drop-dead*, beginning with the mutant as an egg and a pupa and working through a little series of metamorphic stages that are known by the most lyrical name in entomology: *instars*. The poor mutant *drop-*

Benzer is not well known outside science; he is what is sometimes called a scientist's scientist. Here he talks with three of the most celebrated scientists of his century: with Konrad Lorenz (holding drink), a founder of the science of ethology, in Austria; with Richard Feynman (in white shirt), the quantum physicist, at Caltech; and with James Watson, at Cold Spring Harbor Laboratory.

*dead* looks and acts perfectly normal for the first several days of its life, but then it begins to stagger and totter, and suddenly drops dead. Chang has been trying to find out why. This is the kind of laborious work that is required if one is to track the workings of a gene through a fly, and Benzer is happy to be able to delegate some of the bench work to postdocs. Every gene makes its own distinctive protein, and Chang has managed to stain *drop-dead*'s. First he had to purify the *drop-dead* protein and inject it into a rabbit. The rabbit made an antibody that attacked the protein. Then Chang purified the rabbit antibody. Now Chang has a pure *drop-dead* stain, and he can use it like the Alzheimer's stains that Carol Miller is using. But because he is studying the fly he can stain mutants at every age and stage of their lives.

Currently Chang spends most of each workday preparing mutants for the microscope. First he freezes a *drop-dead* fly and slices it into microscopically thin sections, about ten microns thin, on a cryostat. (When they are frozen, flies slice better.) To slice a fly, he freezes some goopy Tissue-tek. The goop turns into a sticky white hill. Then he puts a single fly on the top of the hill and gets it centered. He aims a nozzle at the fly and blasts it with a jet of very cold carbon dioxide gas to keep its walls well frozen. With a fresh razor blade, he whittles away most of the hill's base, leaving the fly lying at the top of a little white Mayan temple. He puts a can over the temple and floods it with more carbon dioxide. The can fills with white shavings, like snow. Then he shakes out the Mayan temple and mounts it in the slicer. When he turns a dial, the slicer spits out a ribbon of dry ice and fly slices. He uses a toy arrow with a rubber suction cup to lift a glass microscope slide and hold it where it will catch the ribbon of dry ice and frozen fly; he manages this step so deftly that the ribbon of shavings adheres to the very center of the microscope slide. Each slide dries for about one hour. Then there is the staining procedure. Then there are rinses and more stains.

One by one, Chang prepares slides of his *drop-dead* mutants, and slowly he works his way down each fly. He stains the antennae, the eyes, and the optical centers, which are right next to the eyes. Even the optic lobe has structure. Some cells are vertical, some transversal; it looks very well organized. Then he stains the brain and the esophagus, which in a fly runs straight through the brain (food on its mind). Chang scans them one by one through the microscope. Wherever and whenever the *drop-dead* gene turns on, at every age and stage of the mutant's life, the gene makes the *drop-dead* protein, Chung's antibody sticks to the protein, and it glows green in the microphotographs that Chang shows Seymour.

"Bright buttons," says Seymour, sipping his tea. Today Chang has been photographing the fly's third instar. There are eight spots on two different planes of focus: six in a bunch, like the six spots on the face of a die, and two spots farther up. They are somewhere in the third instar's ventral ganglion, among a tangle of nerves and tracheal tubes. All of this work is very tentative. It is at the point now where it feels to Benzer like a murder mystery: Who done it, who killed the fly? "Well, you'll have to follow those eight buttons," he says happily, taking the photograph back with him to show off in the Sandwich Shop.

After so many years with *drop-dead* and so many years on the planet, Benzer cannot help identifying with a mutant that walks along normally and then suddenly keels over. Watching it stagger and fall, he often thinks of the victims of Huntington's disease, but the cause of death is completely different. He has no idea if the solution to the *drop-dead* murder mystery will illuminate some aspect of human health. He hopes so, but after all these years he is just curious to know.


THE LATE SENATOR William Proxmire used to give out an annual "Golden Fleece Award" for scientific research so patently irrelevant that everyone who heard of it would have to agree that American taxpayers were being bilked by the National Science Foundation and the nation's universities. Benzer may be the only scientist in the country to have been nominated in the same year for a Golden Fleece Award and a Nobel Prize.

At lunch hour in Harvard's Museum of Comparative Zoology (MCZ), where the old antagonists Richard Lewontin and E. O. Wilson still maintain their laboratories and the views that set them at war twenty years ago, conversations within a few stair lengths of each other are almost as divergent as the views of the old senator and the Royal Swedish Academy of Sciences. When the topic is Benzer, walking from one floor to another at the MCZ is like walking from pole to pole, from one extreme of opinion to the other. "With behavior genetics," as a molecular biologist once put it after praising the clock genes, the celestial beauties of *period* and *timeless*, "you can walk down one corridor at Harvard and walk through all the time zones in the world."

Lewontin is a fine drosophilist, a great contrarian, and a polemicist against much that is central to Western science. "Darwin's theory of evolution by natural selection," he has written, "is obviously nineteenth-century capitalism writ large, and his immersion in the social relations of a rising bourgeoisie had an overwhelming effect on the contents of his theory."

To which the venerable molecular biologist Max Perutz rejoins, "Marxism may be discredited in Eastern Europe, but it still seems to flourish at Harvard."

Lewontin's views on the Human Genome Project and genetic research in general have set him far to one end of the spectrum of debate. He argues that since finding a gene does not tell one what is going on above the gene, a gene is, in itself, worthless information. Benzer and his students would answer: But it is a first step, a way in.

Lewontin has written that the metaphors of science are

filled with the violence, voyeurism, and tumescence of male adolescent fantasy. Scientists "wrestle" with an always female nature, to "wrest from her the truth," or to "reveal her hidden secrets." They make "war" on diseases and "conquer" them. Good science is "hard" science; bad science (like that refuge of so many women, psychology) is "soft" science, and molecular biology, like physics, is characterized by "hard inference." The method of science is largely reductionist, taking Descartes's clock metaphor as a basis for tearing the complex world into small bits and pieces to understand it, much as the archetypical small boy takes apart the real clock to see what makes it tick.

Many of Benzer's students would answer: *Mea culpa.* But look what we found out about the clock!

The clubhouse for the scientists who opened the war on Wilson and sociobiology was Richard Lewontin's Fly Room in Harvard's MCZ. Although Lewontin is loathed and feared by students and eminences of genes and behavior ("He lies! He lies! But don't tell him I said so!"), his Fly Room is decorated like Fly Rooms everywhere. The message BE AFRAID, BE VERY AFRAID blares from the Hollywood poster for the remake of *The Fly,* the same legend that hangs beneath the rifle in Jeff Hall's office. Lewontin's wall clock is decorated with a big paper pair of fly's wings glued to the cinder block with a legend printed beneath it: MOLECULAR CLOCK.

Lewontin is convinced that the Benzer school can find nothing in the fly that matters to the man or woman on the street. "The implication is that if you understand something about the behavior of flies, then you understand something about the behavior of you-know-who," he tells one of his favorite postdocs as they sit down together in the lab for lunch. "That's where the bullshit comes in. Suppose they figure out something about courtship in flies. Maybe they will. They are good scientists. But then what? I don't know what that has to do with courtship in people. So why study it in fruit flies if you are not interested in fruit flies?"

"Always the same story," Lewontin says world-wearily. "You take a simple organism because it is simpler to study, but you factor out everything that's interesting in the process. In terms of *Drosophila*, in terms of the evolution of its courtship, the work is interesting. But it's like saying someone has understood why I eat what I am eating because of smell perception in fruit flies." He smiles dourly. The reason he is eating this particular lunch is not in his genes, Lewontin says, holding it up. "I'm eating it because of my social position in this culture and what's available and what lunch I got to eat when I was a kid." This is a lunch that is typical of his time and his place and his culture but not of his species. "Most people in the world don't eat pizza and cookies."

The postdoc asks Lewontin what he thinks of Tully's spectacular work on learning and memory. "I'm not going to argue," Lewontin says, "that learning and memory in fruit flies has anything to do with learning and memory in us."

Then what did Lewontin teach in the genes-and-behavior course that he himself gave at Harvard some years ago?

"Basic behavior genetics of simple organisms," Lewontin replies. "But I certainly didn't say that mutants in *Drosophila* are why I hate lima beans."

This is one of Lewontin's favorite postdocs; if anyone else were asking these questions, Lewontin might chop off his head. When the young man presses him, Lewontin concedes that it may be possible that we, like the flies, carry mutations that shape our behavior at choice points. "I wouldn't *deny* that," Lewontin says in the tolerant tone with which he might concede that there may be life on other planets in some distant galaxy, to be discovered in some far distant millennium. "But what people really want to know is, why do people want to go around bonking each other on the head? Seriously—there is a suspension of disbelief." Who cares about a clock gene in a fly? "The fact you can find it is nice. But not an earthshaking event." And that there is a homologous gene in human beings is of purely evolutionary interest.

"And *behaviorally*." The postdoc is pushing Lewontin now. If we have clock genes like flies and if variant forms give us variant inner clocks, then the fly work becomes the perfect beginning for the next step, for studies of human clock genes and human behavior, and from there for studies of sex and memory, all working from the genes outward as the Benzer school has done in flies. "So you could pursue a research program using fruit flies as a model," says the postdoc.

"Wait a minute," Lewontin says. Not even a human clock gene is ever going to explain what we are interested in, he says. "I couldn't sleep that night because I had an anxiety attack. Or a fight with my wife. You've got to be very, very careful about carryover. We have very, very complicated neurochemistry. And we don't know how to get at that in fruit flies. We don't know what fruit flies are thinking. We have no idea how to get at that. Maybe if you could get at that . . ."

"Have you ever watched a fly sleeping?" the postdoc insists. "Sitting in its tube?"

Now Lewontin delivers his bottom line. As long as you keep the border very clean and simple and straightforward, he says, there's no problem, no conflict. Flies are flies and people are people. And if there *is* a gene in common between flies and human beings, he says, "then that may be interesting to people like us, who are interested in evolution— but not to anybody else."

The conversation is over. Up and down the table, old friends and colleagues with culturally determined paper-bag lunches are pulling up chairs and sitting down. They are joined by Jonathan Beckwith and Ruth Hubbard, colleagues who are still trying, with Lewontin, to put out the fires of genomania wherever they spring up. The lunch speaker this afternoon is a visiting scholar from Seattle who feels that Western philosophy has been built on the wrong foundations; it should be rebuilt on the proper foundations, the ones laid down by Karl Marx.


IN HIS ANT ROOM in the same museum, E. O. Wilson defends Benzer and his school. "I mean, this is the way it goes," he says in a tone just as theatrically commonsensical as Lewontin's. "Science consists substantially of finding an entry point. And even a small advance following a breakthrough should be regarded as science of the first class. What's unreasonable is then to demand of the people who are doing that very early work—to demand of them that they come up with a whole explanation of everything. That's crazy." Benzer and his students and his students' students have never claimed to explain everything, he says. Nor has anyone else in the molecular study of behavior. "Nonetheless, they should be regarded as people who are beginning to explain behavior.

"As the data come in and the modeling improves and we can face the implications with increasing honesty and lack of fear, my prediction is that the role of evolutionary biology will become increasingly important," Wilson says. "If everything else in biology is the product of evolution, then surely we have to constantly examine and reexamine the human mind and human social behavior as products of evolution."

Lewontin and his group have always argued this way: Until a scientist has proven a particular connection between animals and human beings beyond doubt, they will deny there is such a connection. They will not take the scientific view that the evidence is not in. Instead, they play to the commonsense feeling that there can be no solid connection between animals' instincts and our own. Lewontin's circle sees the attempt to draw such connections as one of the lowest and most base aspects of science since science began: an effort by the elite to repress the masses.

As for Wilson, ever since Lewontin led the attack on him in the 1970s and 1980s, he has been studying the history of science. He sees the effort to analyze the biological origins of human nature, and to unite that body of knowledge with the humanities, as the noblest enterprise in science— the enterprise that has animated science from the beginning and now reaches toward its fulfillment. Sir Francis Bacon considered the shaky state of knowledge in his time and saw "a magnificent structure that has no foundation." There is no other course, Bacon wrote, but "to begin the work anew, and raise or rebuild the sciences, arts, and all human knowledge from a firm and solid basis." Wilson sees Benzer as part of that effort, and he acknowledges the challenge it presents to many sensibilities—essentially the challenge that Darwin threw down in the nineteenth century, made more acute by coming so close to home. What makes the atomic theory of behavior such a radical theory is that it is a theory of relation, or roots—"radical" meaning "of roots." It shows us how we are related closely to our siblings and parents; to every member of our species; to every species from which we have descended; and to all the ancestral forms that began the experiment of life on the planet: all one tree, from the crown to the roots.

The natural sciences have now traveled up from the subatomic particles to the macromolecules, Wilson says, and are beginning to enable us to piece together the cell. We are now facing two great questions: "Question number one: 'Is there any reason to doubt that the same thing can be done for whole organisms and for behavior, even complex behavior?' In my opinion, no, there's no reason to doubt it. The natural sciences have just had an unbroken series of successes doing this. Why should we doubt that we can go on? It may be vastly more difficult, but can it be done? Yes.

"Second question: 'Can it be done from the top down?'" Is there a possibility that major discoveries in mathematics or social theory will allow us to attack the whole subject from the top down, he asks, "and guide the reductionistic and experimental research safely home to harbor? Now I'll give you another opinion: 'No.' I belong to the school that holds that it's gotta be done from the bottom up and that what we need to do is just keep plugging away. These are the exciting areas of research. There have to be Seymour Benzers moving on up into human behavioral genetics, courageously. And as this occurs, then, I think, we're going to see revealed to us principles that will make human behavior and the human mind more comprehensible in ways not yet imaginable. But it has to be done by hard slogging, as they say in the old black spiritual, 'up the rough side of the mountain.' And frankly, that's the kind of—you know, I'm a dirty-fingernail biologist. That's the kind of work I love."


BENZER HIMSELF avoids grand pronouncements. Generations of his students used to smirk with him at a yellowing book advertisement that was taped to the wall in Seymour's Sandwich Shop: "Taking a purely phenomenological approach, Berger examines several closely related concepts: the Second Law of Thermodynamics, organic evolution, ontogenesis, learning, reinforcement, homeostasis, perception, will, consciousness, dreaming, and the Collective Unconscious."

If Benzer ever wrote a book, he would not go on like that poor overreaching Berger. He has always been daring as an experimenter but conservative as a theorizer. Years ago, while sending Delbrück a revised *rII* paper, "a more subdued draft of the rII story" ("It suffers," he wrote cautiously, "from not quite answering any of the $64 questions."), he kibitzed with his mentor about a paper by a newcomer to the phage work. "I had the feeling that he might be squeezing the data a little too hard, and was a little sad at the 'cocktailizing' of theories," Benzer wrote in his letter. "Is it going to get as bad as nuclear physics?"

By now, in Benzer's opinion, the cocktailizing of theories in the study of genes and behavior has gotten much worse than it ever was in nuclear physics. Nevertheless, he is sure of the significance of the genetic dissection of behavior. "Psychology is going to be changed," he says mildly but flatly. In his view the transition will be gradual, as the old guard dies off. "Max Planck said scientists never change. Scientists die and new ones come in. You see that around you all the time. And the very field that is now old hat may ten years down the line get rediscovered. Just like my old mutants.

"I think there may be a tendency that fields change faster now," Benzer says. "New fields are invented more rapidly than the generations turn over. So you automatically have an old guard, a very heavy background of old guard at any one time. I can see molecular biology getting old guard. I can see industry taking over. Molecular biology is going the way chemistry went thirty or forty years ago—hard-core industrial work. Subjects kill themselves off by their success. You say they are successful, but they are no longer at it. They are thriving but not at the cutting edge of intellectual advance. A student can take a course, train to fit in, it's a discipline like every other. Adventurous minds look for somewhere else to go.

"Always academies," Benzer says. "Always the Beaux Arts screaming about the Impressionists and the Impressionists about the Modernists. And the Modern on the Abstract and the Abstract on the Postmodern, and the Post-Post. It's the Feynman thing." One generation sets a question aside, temporarily; and when a new generation comes along, takes it up, and does something with it, the old guard is appalled.

Benzer himself has mixed feelings as he watches the next generation begin what he postponed, approaching human behavior and personality through the genes. But in one way or another he has been working toward this moment and preparing the way toward it through the fly for the last thirty years. He is electrified that now we can begin to see the connections in ourselves, just as we have seen them in the flies. Not long after he and Carol got married, he began asking friends if they wanted to see a snapshot of his new baby boy. The snapshot in his wallet showed twenty-three pairs of human chromosomes. Benzer would mock his own paternal pride by hamming it up as he pointed out the X and the Y. "That's Carol's," he would say, "and this is mine. At least Carol *says* this is mine. . . ." He still finds excuses to show their son's chromosomes at meetings when he talks about the genes-and-behavior stories that are now filling the news.

His friends laugh. "Only a geneticist."

CHAPTER EIGHTEEN

# The Knot of Our Condition

*The knot of our condition was twisted and turned in that abyss.*

—BLAISE PASCAL,
*Pensées*

BENZER KEEPS a clipping file of genes-and-behavior headlines so that as they are discredited he can use them in his lectures as cautionary tales. In the three decades between 1965 and 1995, studies were announced— often with great fanfare—linking human genes and violence, reading disabilities, manic depression, psychosis, alcoholism, autism, drug addiction, gambling addiction, attention deficit disorder, posttraumatic stress disorder, and Tourette's syndrome. Every one of these studies had to be retracted.

Today, with the tools of molecular biology growing more sophisticated and the maps of the human genome filling in, Benzer thinks it is possible to do good work at last. Skeptics like Lewontin say the good work will never come, that this whole field will be remembered someday with the same contempt we now lavish on Galton's eugenics. But Benzer thinks that thousands of solid links between genes and human behavior will be discovered over the next several decades. He is eager to read these stories, and like everyone else he is particularly hungry for information about the traits that have shaped his own life. He thinks he is probably a clock mutant, and in the middle of the night he sometimes marvels at the gift that this one mutation has given him, a lifetime of solitude in the laboratory. Human clock genes are now being cloned, sequenced, injected into the eggs of mice, and dissected by the techniques that Benzer and his students pioneered in the fly.

Benzer also wonders if he is a thermostat mutant. His lab coat, shirt, and sweater do not always keep him warm in Church Hall, although everyone else in his lab wears T-shirts. These extra layers are not a sign of age; they are a sign of Benzer. "My fingers are cold," he has told people all his life. "Feel them. I'm ten degrees off everybody else." Even forty years ago, camping in the desert with Dotty and the Delbrücks, Benzer wore two sweaters, two pairs of pajamas, and something around his neck. Dawns in the desert were a daily double whammy because they were so cold and so *early*. The Delbrück family had a saying: "More tired than Seymour at the Grand Canyon." Whatever the cause of it—probably poor circulation—it is a single brush stroke that has shaped his life. One reason Benzer works at Caltech and not at Harvard, which has made and lost five bids for him over the years, is his dislike of snow.

This is the way we define ourselves. We single out a few traits from all the tens of thousands, a few traits that vary a great deal from everyone else's, and we watch their effects at the choice points the way Benzer does with his flies. At Benzer's seventieth birthday party, Francis Crick told a few stories about a sabbatical year Benzer spent with him at Cambridge in the late 1950s. Crick and Benzer sat in the same tower room at the Cavendish Laboratory where Crick and Watson had discovered the double helix, fiddling with the same rods and cutout pieces of tin with which Watson and Crick had built the first model. "We gradually got used to Seymour's habits," Crick remembered at the birthday party. "Not getting in too *early* . . ." Benzer always claims Crick's tower was drafty, but Crick says, "Well . . ." No one else in the lab ever noticed any drafts. "We surmised, by sort of inspection, that Seymour was wearing more than one sweater, probably two or three, I think. And I have heard that on occasion he even wears two pairs of socks. Seymour," Crick announced, enjoying himself, "I've come to the conclusion, for reasons that I'll mention later—perhaps in your style of work—that you must have a very low metabolic rate, and this accounts for what might be called both the *exo*-insulation and the *endo*-insulation."

That was an insult worthy of a genius. Crick was implying that a single genetic flaw explains four of Benzer's most colorful traits: one, his outer layers; two, his inner layers (because there have been years when Benzer has cast a round shadow even without the sweaters); three, his late hour of rising; and four, his wittily simple experiments. Crick likes to hint that all four of these traits of Benzer's derive from just one defect: Benzer is lazy to the core, and he invents his cut-to-the-chase experiments out of sheer indolence. This is Crick's favorite dig at Benzer. It is the way he describes Benzer in his memoirs: "always one to avoid unnecessary work." But then this is also Benzer's favorite dig at Crick. Benzer simply can't understand how everyone in Crick's laboratory spent the morning drinking coffee, the afternoon drinking tea, and then got called to Stockholm for a Nobel Prize. "I don't know," Benzer said at his birthday party, parrying Crick, "it wasn't clear at all when they did their work. At Easter vacation, all the gas was turned off in the laboratories. And at night, to get in, you had to wake up the concierge to let you in through the gate. So I still don't quite understand that miracle."

Because of his eccentric thermostat, Benzer took a personal interest in the work of one of his recent postdocs, Omer Sayeed, from Pakistan, who looked for thermostat mutants by putting flies in a clear plastic tube that sat on an aluminum slab. One end of the slab was hot, and the other end was cold. Wild-type flies always chose the middle of the slab, around 24° C. That seemed to be their Pasadena. Sayeed tried raising flies in a hot room and in a cold room, but when he gave them the chance the flies still chose Pasadena. The preference is innate, and that fascinates Benzer.

Sayeed also used the slab to test some of Benzer's classic mutants, including one of the very first eccentrics that Benzer discovered in his countercurrent experiment, *SB-8* (meaning *Seymour Benzer's Eighth*), a mutant fly that does not go to the light. *SB-8* turned out to be a thermostat mutant too. It did not prefer any particular piece of real estate on Sayeed's slab, even if he made the slab icy at one end and infernal at the other. The fly seems to be thermo-blind. Sayeed and Benzer have renamed it *bizarre*.


DEAN HAMER, at the National Institutes of Health, is the most prominent molecular biologist to enter the field that Benzer pioneered, and to look at human beings. Hamer is gay, and in his first study of genes and behavior in the early 1990s he decided to study why some human beings are attracted to members of the same sex, while most human beings are attracted to members of the opposite sex. Hamer thought of this as a relatively clear-cut and dramatic behavioral variation with which to start his study, just as Benzer started by studying flies that turn away from light and Jeff Hall started by studying male flies that court other males.

Of course, it is harder to study a man's choices than a fly's. How much of the difference in sexual orientation between two American men is imposed by the way they see themselves and the way they try to behave in their culture? There are still many psychologists who argue that human sexual orientation is determined more by culture than by biology, more by nurture than by nature, which is less of an issue with *fruitless*. As a young lawyer, Abraham Lincoln shared a bed for two years with a roommate in Springfield, Illinois. Historians now argue about what that meant and whether Lincoln was homosexual. In *The Invention of Heterosexuality*, one American historian (also gay) argues that the very idea that most men are attracted to women and most women to men is a social invention.

But Hamer feels it is reasonable to assume that much of the difference is inborn. Many psychologists agree, and many gay men describe that as their subjective experience, in the tones of the Roman poet Horace: "Drive out nature with a pitchfork, she'll always come back." Or Voltaire: "We perfect, we smoothe down, we hide what nature has placed in us, but we put nothing there ourselves." Twin studies in the early 1990s showed that among nonidentical twins, if one is gay the chance of the other being gay as well is about 25 percent. But with identical twins, if one is gay the chance of the other being gay is 50 percent. These findings suggest that genes help to shape the variation in sexual preference. At the same time, if one identical twin is gay, there is a 50 percent chance the other will be straight, so it is also clear that genes do not decide sexual orientation the way *white* and *fruitless* decide eye color and sexual habits in flies. The neuroanatomist Simon LeVay (also gay) believes he has found anatomical differences in the brains of gay and straight men. Although his findings and their implications are controversial, LeVay has reported differences in the hypothalamus, differences as marked as those that other investigators have found there in men and women.

Hamer recruited study subjects through outpatient HIV clinics and gay men's organizations in Washington. He took blood samples from each of his volunteers and administered various personality tests. He also did a standard pedigree study of each volunteer, looking for homosexual relatives in each family tree. Hamer was intrigued to see that the gay men in his study were more likely to have gay uncles and gay cousins on their mother's side than their father's. Every biologist since Morgan would know what that suggested: that there might be a link between the trait and the X chromosome. Since a man has only one X chromosome and he gets it from his mother, any trait linked to the X will pass down through the mother's side of the family.

If a gene on the X chromosome makes a man more likely to be homosexual, two homosexual brothers should be likely to share that gene and also some of the genes around it. This is the same mapping principle that Sturtevant hit upon in Morgan's Fly Room. Hamer checked a series of twenty-two markers that span the X chromosome. By now he had his choice of computer programs to crunch the numbers for him (he used LINKAGE 5.1). The program pointed to a link between the homosexuals in his group and a marker at the far end of the long arm of the X, at a site called Xq28.

From his data, Hamer could not tell what the gene might be, how many male homosexuals in the population at large might carry the allele at Xq28, or what portion of their sexual orientation was influenced by that allele. Hamer could say only that somewhere within about four million base pairs on the tip of the long arm of the X chromosome there might be a gene that might somehow relate to the sexual orientation of the men in his particular study. In other words, compared to the kind of work that had been done for decades with flies, the finding was tentative, and, being a careful molecular biologist, Hamer presented the finding to his colleagues that way.

But genes, behavior, and homosexuality are such charged subjects that Hamer's story caused a national sensation. Within days of the announcement, many gay men were buying a T-shirt in gay bookstores: "Xq28—Thanks for the genes, Mom." At the same time, gay activists denounced the work; they were afraid that the suggestion that homosexuality is in the genes might someday lead another Hitler to attempt another Final Solution, or lead millions of parents to use a prenatal diagnostic kit. The "gay gene" story provoked furious controversy both in the press and in the scientific establishment. A young postdoc in Hamer's laboratory who had helped map the gene to Xq28 accused Hamer of picking and choosing which of his data to report. This was a serious charge. Hamer's colleagues at NIH began a confidential inquiry, and so did the Office of Research Integrity in the Department of Health and Human Services. After news of the ethics investigation broke in the *Chicago Tribune,* Hamer sent a note to *Science* by e-mail defending himself and saying that he doubted there would be so much controversy if he were working on any topic other than homosexuality. He was cleared in 1996, and all charges were dropped. Meanwhile, a study in Canada found no evidence of the linkage that Hamer had seen—not even a link to the X chromosome, much less the tip of the long arm of the X chromosome. But that study was never published.

In the middle of these storms, Hamer enjoyed hearing about a discovery by two colleagues at NIH, Ward Odenwald and Shang-Ding Zhang.

They were looking at the development of the fly's nervous system, studying a gene called *pollux* (which acts in concert with a gene called *castor*). To find out what *pollux* does, they had made a DNA transformation cocktail with *pollux* and a heat-shock promoter, so that the gene would turn on only when they turned up the heat. All this was standard procedure by now. Also following standard procedure, they used the early embryo of a white-eyed fly and added the normal allele of the gene *white* to the DNA transformation cocktail, so that they could see at a glance which flies came out of their eggs transformed. A fly that popped out with red eyes would carry the gene *pollux*. When Odenwald and Zhang watched these flies in a warm room, they were surprised to see the male flies begin dancing around and around in circles on the walls of the fly bottles.

After a year's study, Odenwald and Zhang decided that the gene that had made the difference for the flies was *white*, the gene that started modern genetics. They could make fruit flies chain just by injecting the normal allele of *white* into the eggs and turning up the heat. They even speculated in their paper that their gene might turn out to be a clue to homosexual behavior in human beings, which was a naive leap; and again the story attracted national press attention. *Time* magazine ran the headline "Search for a Gay Gene," with the headline wreathed in a circle of chaining male flies.

The finding has since been replicated at Yale, but the basis for the effect remains unknown. Until it is explained, the leap from *white* to human beings is at best premature, as Jeff Hall explained at the time of the discovery to everyone within earshot. "It's completely silly," Hall told a reporter from *Science News*. "Nobody between now and doomsday will think *white* is going to have anything to do with behavior in *mammals*. The chance of this is one over the number of neutrons in the universe."

Of course, there is a lesson in *white* for human beings and for the human future. By now, *white* is one of the best-known genes on earth. It is the gene that put genes on the map, the cornerstone of modern genetics. Drosophilists have now been working with it in Fly Rooms all over the planet for most of a century. And *white* is also the gene that started the whole century of talk of "a gene for." It always seemed the simplest possible model of a gene linked with a trait. That such a gene can cause such complicated and unforeseen behavior when injected into a fly is a cautionary tale for those who may begin in the next few years to think about injecting genes into the eggs of human beings, even genes linked with apparently simple and innocuous traits like blond hair or blue eyes.

This is why so many drosophilists stay away from the human stories. "I wouldn't touch that one with a barge pole," they tell one another when they see headlines about Hamer.

"Flies have no political constituency."

"It's a sobering thought," Tully says, when he considers—as he must—that his work on *creb* may someday lead to attempts at genetic engineering, attempts on the human brain. For him the apolitical approach of the last generation is no longer possible, if it was ever tenable. For a molecular biologist of his generation the twentieth century teaches the impossibility of pure research. "What went through Einstein's head when he saw *E* equals *mc* squared?" Tully asks. " 'Shit, we can blow up the planet'? Did he say that? I'd suspect in his dark hours he knew it would be abused. We have the same phenomenon here with this enhanced memory. We can see it now, and we know it's real. And that really kind of brings in a new day." The genetic dissection of behavior has concrete implications now that once seemed like science fiction. "Now it is science fact, like fission. There is a potential here for serious abuse."

Tully wonders what the military would do if it got hold of drugs for memory enhancement. "Think about it. A perfect drug for the CIA." Send in agents, take a memory enhancement pill, the agents have brilliant memories during the operation—and then they lose it all afterward. "And then you weren't there. Perfect. You know? Think about the pressure of a general who has thirty minutes to communicate a data-rich conversation of specifics of bombing missions to a group of pilots before going off to drop bombs. Do you think he'd cram, then take a memory enhancer? They'd be champing at the bit for drugs that could modulate memory in that fashion. And yet that's not what we want them for. I'm a pacifist. I would hate to see this understanding perfected for the art of war, for all the covert and overt atrocities that humans push over on each other. But it's possible. You could go science fiction. What would it be like if a child popped a memory enhancement pill every day before school? What would that child's head be like after twelve years of education? What would the child accomplish with that store of information? That's an interesting question to ponder. And would it even work? Can the brain deal with it? Is the capacity there to deal with what we're imagining could be produced? We don't know."

Tully thinks of clocks too: "Maybe arrhythmic mutants lead to depression. Take the drug, you cycle. Well, does that mean you could put it in the water supply of the Iraqis?" In other words, even an apolitical gene like *period* may lead to medicines and weapons. "Could it work?

I don't know. What would industry pay for a drug that could easily set and maintain the clock for swing shifts in the plant? Is that what we want to do?"

One of the most fascinating applications of his work, he thinks, would be a drug to block memory of trauma. He could use the off switch as easily as the on switch and make an amnesia pill: "The perfect treatment. Cutting it off at its source." That would be even better than erasing memory of trauma: preventing it from being written down in the first place. "That could be the first and maybe always the best outcome of what we're doing. Improve conditions for those who experience some really bad, sad, powerful thing. So do 'em a favor and wipe it off. Then they won't suffer from the memory of it.

"Then again, I wake up in the middle of the night and say, 'Yeah, but would I be who I am without suffering?' That's a tough one. Thank God I don't have to answer it. I just play with flies."

This is why Benzer is just as happy to study the eye of the fly instead of the behavior of the fly and to set the world of politics at a distance. It is a choice that fits the cricket-in-a-cage approach to science that he learned from Delbrück and his generation. For Morgan and most of his Raiders too, it would have been undignified and inappropriate to get involved in campaigning against health fairs and eugenics programs. To follow pure science was the Arrowsmith ideal.

So Benzer follows his curiosity wherever it leads inside the fly. To an outsider that might sound confining. But to a drosophilist today the scope is infinite. As the century closes, there are six thousand drosophilists around the world, and their number is growing by 20 or 30 percent every year. Flies have turned out to be far more like us than anyone imagined in the 1960s, when Benzer (shocking his friends) turned back to the fly. Faster and faster, drosophilists add genes to their Web site FlyBase. The naming continues to be more whimsical and irreverent than physicists' names for new elements, which tend toward the monumentally serious. FlyBase includes descriptions of all of the latest genes. In this way, drosophilists are continuing another tradition that Morgan and his Raiders started: sharing information as they get it, and not hoarding it as many other students of genes do. A recent story in the *New York Times* about the return of the fly ("NOW PLAYING AT A NEARBY LAB: 'REVENGE OF THE FLY PEOPLE'") began with a list of just a few of the stranger names: "*Godzilla, genitalless, gut feeling, gouty legs, goliath, gooseberry distal, ghost, glisten, gang-of-three.*"

"Every single biological phenomenon on the face of the Earth or in the universe is studied now in *Drosophila*," says Jeff Hall. "We're not drosophilists anymore, we're biologists who happen to use *Drosophila*. I mean, *Drosophila* meetings now are a joke. They're about every aspect of biology under the sun."

Hall is still angry that his old boss Benzer ranges so widely in the fly— while neglecting behavior. "Benzer is an antidetective," Hall says. "He doesn't ever figure out anything. He's not interested. Once problems get *intense*"—once a crowd of people converge on them—"he loses interest and *drops them* and starts looking for new things. In fact, if you look at the array of subjects on which he continues to publish, one sees an explorer who is rattling around the biological landscape like a *superball!*"

IN JANUARY 1996, Hamer got a call from Israel. A team of molecular geneticists there was doing a study they thought might interest him. They had taken blood samples from a group of subjects and administered a personality questionnaire designed to measure four domains of temperament: novelty seeking, harm avoidance, reward dependence, and persistence, four traits that seem to a number of psychologists and behavior geneticists to be partly inherited. The novelty-seeking scale of the questionnaire tries to sort people into those who are more "impulsive, exploratory, fickle, excitable, quick-tempered and extravagant" and those who are more "reflective, rigid, loyal, stoic, slow-tempered and frugal." The Israeli investigators found that those who scored higher than average in novelty seeking were also more likely than average to carry a certain variant form of the gene for one of their dopamine receptors.

Dopamine receptors are famous in psychopharmacology because they are primary targets for drugs that are used to treat many neurological diseases, including Parkinson's and schizophrenia. Pharmacologists and psychiatrists often prescribe a drug called clozapine for schizophrenics who have not responded well to other treatments. Clozapine binds with peculiar affinity to one particular dopamine receptor, D4. The repeats in the D4 dopamine receptor can lead to differences in its affinity with drugs, at least in laboratory tests. The gene is expressed in the frontal cortex, midbrain, amygdala, and medulla of monkeys, parts of the brain that are linked with cognition and emotional behavior. Amphetamines, cocaine, and alcohol are thought to change our mood by altering dopamine levels; so do antipsychotic drugs such as clozapine or haloperidol.

The gene for D4 is on the short arm of chromosome 11, and the gene contains repeats. In some of us a run of forty-eight base pairs within this gene is repeated twofold; in some, fourfold; in others, sevenfold.

While the Israeli team was studying the DNA of their volunteers in the Negev desert, behavior geneticists in England and in Boulder, Colorado, looked at what behavior geneticists working with mice call "emotionality," or sometimes "reactivity." When a mouse is placed in an apparatus they call an "open field"—a brightly lit white circular arena, a sort of spotlit stage—one mouse will spend most of its time exploring the stage, while another will spend most of its time keeping very still and defecating. The mice also behave in character when they find themselves in the dark arms of a Y maze. Their behavior can be predicted from their lineage. The investigators crossed mice that explored the stage with mice that fled the stage, tested their grandchildren, and looked at the DNA of the most extreme mice at each end of the scale. Then they entered all the genetic data in a computer program called MAPMAKER and found at least three loci, on murine chromosomes 1, 12, and 15, that seemed to be linked to a mouse's emotionality.

In the Israeli sample, most subjects had either four or seven repeats. The higher the subjects' ratings in novelty seeking, the more likely they were to have the sevenfold repeat. So Hamer and some of his colleagues at NIH tested the blood samples they had already collected in their study of what has become known as the "gay gene," together with other samples they collected from local college students. Then they re-sorted their subjects into two groups. One group had short alleles with two to five repeats; the other group had long alleles with six to eight repeats. When they checked the personality tests, they found that the long group scored higher on warmth, excitement seeking, and positive emotions. The long group also scored lower on conscientiousness; specifically, on a facet of conscientiousness that the test makers called deliberation.

In Church Hall, Benzer combed through these new studies with the same mixed curiosity and asperity with which he had looked over Hamer's claims for Xq28. Benzer has always felt that his own key trait is curiosity. In the hall outside his workroom, he keeps six spring-loaded steel file drawers full of maps: of Paris, Cambridge, Delbrück's deserts, everywhere Benzer has ever been and hopes to explore again ("I don't know, am I going too far? A map is a wonderful thing."). In the same mood he now spends whole nights trolling the World Wide Web for its bizarreries. But he has added Hamer's novelty-seeking story, which has not yet been solidly confirmed, to his clipping file. He thinks the finding may hold up, but again it has been absurdly overblown in the press. He mistrusts those multiple-choice personality questionnaires ("I think they're scandalous") and he suspects that the gene is a smaller beginning than the media hooplah suggested. According to one recent twin study, novelty seeking is about 40 percent heritable. By Hamer's calculations, the dopamine-4 receptor gene accounts for about a tenth of that. At best the D4 gene would account for about 4 percent of the trait. So why call it "the novelty gene"?

But in poetic if not in scientific terms, the name does have appeal, as the lead of the front-page story in the *New York Times* observed: "Maybe it is appropriate that the first gene that scientists have found linked to an ordinary human personality trait is a gene involved in the search for new things."

LATE IN 1996, Dean Hamer and another group of investigators announced that they had found a link between a human gene and the pursuit of happiness. This time they focused on a gene that codes for a protein that helps nerve cells to recycle the neurotransmitter serotonin. In human beings a certain transporter of serotonin, 5-HTT, is expressed by a single gene on chromosome 17. Hamer and his collaborators found a variation of the gene's coding region about one thousand base pairs upstream, in a place that controls the gene's transcription. There are repeats in the DNA here, and again most people in their sample fall into one of two groups: a group with a short form of the gene and a group with a long form.

Hamer found that in his sample of volunteers those who had two copies of the short form of the gene scored higher in neuroticism than those who had two copies of the long form. The variation in the gene showed no significant connection to variations in the other personality characteristics: extroversion, openness, conscientiousness, and agreeableness. And as with dopamine, there is strong reason to believe that serotonin has a strong effect on mood and temperament. Drugs that inhibit the uptake of serotonin are often prescribed in the treatment of anxiety and depression. Changes in the transmission of serotonin cause anxiety in both animals and human beings.

Again Benzer and his students were skeptical and waited to see if the findings would be replicated. But the press and Hamer himself greeted the discovery ebulliently. When Hamer's computer program first found the link, he told his friends, "We found a happiness gene!—I shouldn't call it that." The day the study came out, he was quoted on the front page of the *Philadelphia Inquirer:* "Everybody will be happier."

Again the overinterpreting and overreporting in the press made the fly people and the mouse people glad they were staying with the fly and the mouse. Those come close enough to home. One brown mouse gives birth to a litter of pups. She nurses them, and she herds them back into her nest when they stray. Another virtually identical brown mouse gives birth to a virtually identical litter. But she never nurses them. She lets the pups wander farther and farther from her nest in the cedar shavings at the bottom of the cage, and almost all of them die.

One white mouse snuggles for hours with the other mice in its cage, trimming their whiskers and letting them trim his. Another virtually identical mouse keeps to itself at the far side of the cage. Its bed in the cedar shavings is unmade and unfluffed, and its whiskers are untrimmed.

One maggot, when it crawls to a crumb, always takes one or two bites and crawls on to the next crumb. Another virtually identical fruit fly maggot arrives at a crumb, settles down, and eats every bit before moving on toward the next crumb.

The difference between the first mouse and the second, the mother superior and the mother inferior, is that one of them has a normal set of mouse genes and the other is missing a gene called *fosB*. The difference between the well-trimmed mouse and the unkempt mouse is that the second mouse has a problem in a gene called *disheveled*. The difference between the roving maggot and the sitting maggot is a single letter of genetic code in a fruit fly gene called *foraging*, also known as *dgk2*, at map position 24A3–C5 on the left arm of the second chromosome.

The laboratories that engineered the mice are at Harvard Medical School and the U.S. National Human Genome Research Institute in Washington. The laboratories that created the roving and sitting maggots are in the open air, because this is a natural variation. Roving and sitting maggots are found wherever fruit fly larvae wriggle out of fruit fly eggs, which is virtually every temperate spot on the planet. Every fruit fly has to creep on the face of the earth for a few days as a maggot before it can metamorphose and take to the air. Apparently, among maggots both rovers and sitters are viable personality types.

Every human being also has a copy of the mouse gene *disheveled*, the gene that is damaged in that unbarbered and unsocial mouse. Every fruit fly has a copy of *disheveled* too. In fact, like so many thousands of genes that now interest biologists, *disheveled* was first discovered in fruit flies.

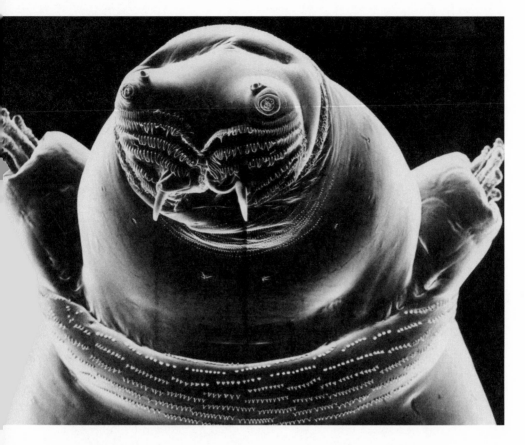

*Portrait of a maggot, in a scanning electron micrograph. Some fruit-fly maggots are rovers, nibbling here and there and moving on; other maggots are sitters, eating every bite before moving on. The difference between the rover and the sitter is a single letter of genetic code in a gene called* foraging.

Drosophilists named it *disheveled* because a fly with a disordered form of that particular gene always pops out of the egg with his chest hairs in disarray.

WE LOOK AROUND the family table and see some fragments of behavior that seem to come out of nowhere, other fragments that we recognize instantly. Often it is ourselves we recognize. We catch glimpses of the way we chew or talk, laugh or frown, right down to the way we pour from a pot or sip from a cup. The secret faces of our inner lives glance back at us from the fracturing ripples of the gene pool. We also catch glimpses of ourselves in the faces of our animals, as if they, too, are reflected in the same wavering pool. These resemblances will fascinate the last human generation as they fascinated the first.

A computer operator from the south of France goes back to his ancestral village in Ethiopia. His family left Africa years before he was born. But when he meets his grandfather, who is the chief of the village of Shembe, three hundred miles from Addis Ababa, he sees that they not only look alike, they look at the world alike and move through the world alike. After the computer operator goes home to France, he learns that his grandfather has changed his will and named him the next chief of Shembe.

A teacher from Texas goes back to his ancestral village in Scotland. His grandparents are dead, but he sits down to tea with a great-aunt. He offers to pour, and as he tips the pot his great-aunt gives a little cry: "Oh, my God, your granny! It's in your hands!"

A mother calls a therapist to talk about her son. The boy has just turned fifteen, and he is acting as loutish as his father, a man she threw out of the house a little more than fifteen years ago, a man the boy has hardly met. Does she have a gift for turning men into louts, or is her son his father coming back?

A mother in Manhattan watches her son's face as he sleeps. She left his father in Paris soon after the boy was born. Her son hardly knows him. But more and more often, even when he is dreaming on his pillow, she seems to see in his face expressions that remind her of his father, expressions that seem to her impossibly, indefinably French.

All these anecdotes point in the same direction as the celebrated studies of identical twins raised apart, including the twins who are both gunsmith hobbyists; the twins who are both raconteurs; the twins who are both hysterical gigglers; the twins who can enter the water at the beach only by backing up, timidly, "and then only up to their knees." Of course, there are anecdotes that point in the opposite direction: the young girl who looks and acts like neither of her parents; the adopted boy who grows up to walk, talk, think, and laugh exactly like his adopted father.

"I think genes and behavior are such a headline item," Benzer says, thinking of his clipping file: GENE DISCOVERED FOR BEDWETTING. GENE TIED TO LOVE OF NEW THRILLS. "But the trouble is, when you go look at the data, they are often really fragmentary." He sees dubious measurements and marginal correlations. "Much as I believe in genes and behavior, the idea has caught on too much. It's become an idea of complete destiny. I think that's wrong. Genes are not always expressed. Even if you work with fruit flies, you see that genes are not always expressed." We each carry many genes we never express. The likelihood that we will express a gene we carry is called the gene's penetrance. Penetrance is not the same for each gene. "Look at the bible," says Benzer—meaning the drosophilist's bible, *The Genome of Drosophila Melanogaster*, a book that lists every fruit-fly gene ever found since *white*, and rates the penetrance of thousands of them. "You can have a gene with ten percent penetrance, or five percent, or one percent. So just having that gene doesn't mean you'll show that phenotype. Expression depends on a myriad of chemical reactions. And that's not generally understood. People think if you have the gene, your fate is sealed."

Benzer is sure that when the picture of genes and behavior begins to fill in, there will be no such thing as "the gay gene" or "the curiosity gene" or "the happiness gene." All these traits will prove to be at least as complicated as a fly's tendency to move toward light—and Benzer now knows hundreds of genes that affect that single trait. Students of genes and behavior will dissect vast complexes and constellations of genes that work together, as in the clockwork in the fly.

But as the science he helped to start comes closer and closer to home, he sees patterns and questions everywhere. He visits his grandson at his high school during lunch break, and he thinks: What a field for study. His grandson says that every lunch break, the same students stay inside and the same students go outside. Outside, there are the students who lean on the cars, the students who sit around by the bikes, and the ones by the flagpole. Each group has its own attitudes and makes its own moves at the choice points. Benzer is sure that behind these choice points and behind all the schoolhouse culture that surrounds them, there must be a thousand and one differences in the genes. The choices may be too complicated to dissect at the moment, but the influence of the genes is real and ever present. "It's not random," he says. "None of it is random."

He daydreams now in the middle of the night about simple traits that one might dissect soon. Sometimes he remembers Galton's old idea about the instinctive dread of blood. Benzer once had a graduate student who was an extreme case. He would faint at the sight of blood, even the mention of it. He passed out once at the Faculty Club, and once at Benzer's house, too. People would forget themselves as they stood around talking shop and munching adventurous hors d'oeuvres in the living room, and there he would go again. "Try to catch him! A real phenomenon."

Then there is the drinking question. When Benzer watched his crews of postdocs back in the bacchanalian sixties and seventies, he used to remember an old Yiddish song from Brooklyn. A Jew goes into a bar, he drinks a thimbleful of wine. A goy goes into a bar, he drinks a barrelful of wine. And the chorus:

> *Drunk he is*
> *drink he must*
> *because he is a goy—*
> *Hey!*

Jeff Hall, who is half Irish, likes to make an ironical toast when he hoists another brown bottle at midnight: "Those Irish alleles!"

The next generation of molecular biologists is trying to study those choices now, and there are signs that this trait, complex as it is, may be illuminated by studies of genes. Lee Silver of Princeton University is another molecular biologist of Hamer's generation who is moving into the study of genes and behavior, and alcoholism is one of the traits that Silver is studying now. He works with mice, and he finds the current possibilities for research in murine genes and behavior so exciting that he often wishes he could extricate himself from every other project in his lab and do nothing else.

Of course, for studies of behavior as complicated as alcoholism, a mouse makes a somewhat problematic model, as Silver himself points out. A mouse will never say, "Gee, I'd like another drink, but I guess I shouldn't." A mouse will just take that drink. On the other hand, this too makes the mouse useful: all those layers of willpower, experience, education, and nurture do not come into play. In fact, looking for links between genes and behavior is so straightforward with mice that Silver lets his undergraduates do most of the work. Not long ago, one of them designed an experiment in which she offered inbred strains of mice two spigots, one for water and one for alcohol (10 percent ethanol, about as strong as Chardonnay). An inbred mouse strain known as C57BL/6 will drink three quarters or more of its liquid from the alcohol spigot. A second inbred strain, DBA/2, will drink almost none—less than one hundredth of its liquid diet will come from the alcohol spigot. A DBA/2 mouse drinks so little alcohol that it is likely to take no more than a single small taste from the sipper tube and never go back.

A senior of Silver's, Justine Jaggard, crossed these alcoholic mice and teetotaler mice. Then she crossed the children with teetotalers. Some of the grandchildren drank a great deal of alcohol. Some drank almost none. Jaggard tested the DNA of the mice for a large number of markers that she knew to differ in the alcoholic and the teetotaler strains. Now she could see which of these markers were most often found in the alcoholic mice. Those markers had to lie next to or near to the genes that made the difference in their behavior.

One night in June, just before the end of her senior year, Jaggard found a locus on mouse chromosome 2 that seemed to predispose male mice to alcoholism, and a locus on chromosome 11 that seemed to predispose female mice to alcoholism. The gene in the female mice seemed to account for roughly a fifth of the variance in their drinking patterns. She called Silver first thing the next day: "I got an awesome result last night at three o'clock in the morning. I know. I can't believe this actually worked. I'm so excited. Hooh! But anyway. Well, it's real! I'm totally, totally excited. But there's no mistake!"

Today students of Silver's are crossing aggressive and passive strains of mice; mice that are subject to seizures when they hear loud and high-pitched noises and mice that are immune to those same noises; mice that are monogamous and mice that are polygamous. With each of these traits Silver expects to begin finding complexes of interacting genes and dissecting those complexes, while he looks for corresponding genes in human beings. He also has a graduate student who is making mouse mosaics to help trace the fine-grained differences of their behavior to their brains. They will engineer a mouse so that half its cells are male and half are female: random bits and blotches of maleness and femaleness, from the fur to the brain. Then they will test these mice and see how their behavior varies, depending on which part of the brain inherits which genes. "It's a fantastic idea we came up with together," Silver says. "We get this from *Drosophila* gynandromorphs. Conceptually, it's the same thing: genetic dissection as opposed to surgical dissection."

Like most biologists in this exploding field, Silver speaks enthusiastically of the genetic dissection of behavior without thinking of the source of the phrase. "A lot of this comes from Seymour Benzer's vision," Silver acknowledges. "In the back of my mind, that's where it's coming from, even if we don't express it. He was the one. And from that vision it just goes on everywhere."

# CHAPTER NINETEEN

---

# Pickett's Charge

*Human knowledge will be erased from the world's archives before we possess the last word that a gnat has to say to us.*

—HENRI FABRE

OUT ON THE BATTLEFIELD at Gettysburg, more than one hundred molecular biologists gather again and again in great semicircles around their guide, who is shouting out the history of the fight through a white bullhorn in hectic italics, like the hero of a comic book: "This is the *Wheat Field,* and it's a *shame* they don't keep it planted in *wheat*! They keep the Corn Field at *Antietam* planted in *corn.* As well they *should!*"

It is a fine fall afternoon, and the molecular biologists and their families tromp along in broken ranks, chattering among themselves in English, Japanese, Chinese, French, and German. They cannot help marveling at the supernatural quantities of military information that their guide has stored in his head, although when one of the graduate students in the front ranks hurries ahead to compliment the guide himself, he lowers his bullhorn for a moment and replies, "If an idiot savant can be said to *know.*"

The symbol on their ID tags is the Princeton University heraldic shield, orange and black, with little legs added to the bottom of the shield to make it look like a virus: a bacteriophage. This is Princeton's Department of Molecular Biology, the university's wealthiest and fastest-growing department. Their laboratory on the edge of the Princeton campus was designed by Robert Venturi with architectural allusions to the Doge's Palace in Venice. When the heads of the department chose Gettysburg for this year's annual retreat, they asked a colleague at Princeton, the historian James McPherson, author of *Battle Cry of Freedom,* if he would be willing to show them around. McPherson told them to call Jeff Hall. "Hall knows more about the Gettysburg battlefield than I do," McPherson said, "and Hall is a biologist besides."

Last night in the Robert E. Lee Room of the Gettysburg Ramada Inn, the drosophilist Eric Wieschaus, who prepared for this retreat by reading *The Blue and the Gray,* sat with Hall, talking about Gettysburg and about genes and behavior. Late in the evening, Wieschaus was grappling with the paradoxes of Hall's field—and grappling with his own hair, shoulders, and torso as he struggled for precise thought. Hall announced to the table, "We've just watched Eric Wieschaus wrestle himself to the ground. As he does often."

Wieschaus laughed. "Even in the middle of talks," he said. "Even in the middle of major talks." Twelve months later he would be getting his dawn call from Stockholm.

Even out here on the battlefield, most of the molecular biologists in the ranks are talking molecular biology. Their science is racing ahead so fast that they rarely take time out for an afternoon like this or for a look backward at the history of their own battlefields. Not long ago in Pasadena, Seymour Benzer's Korean postdoc brought him a petri dish. He had injected the fly gene *drop-dead* into the bacteria in the dish, and now the bacteria were dropping dead. Benzer studied the plate with amusement. He knew that his postdoc was disappointed. The postdoc was trying to make the bacteria express the *drop-dead* gene so he could study the *drop-dead* protein; instead, the bacteria were simply dying. "I like the idea of making bacteria drop dead," Benzer said. "I used to make them drop dead with phage."

"How do you do that?" asked the postdoc.

"They eat bacteria. The phage I worked with—" Then it struck Benzer. "You don't know that? Oh, my God!" he cried with good-natured despair. "So the whole phage literature passed you by?"

With the science accelerating so rapidly, all of the founders feel like ghosts standing in their own fields. There are postdocs in an institute named for Delbrück in Germany who have no idea what Delbrück did. Postdocs at meetings in Cold Spring Harbor see Watson striding by and exclaim out loud, "He's still alive?"

"Molecular biology has no history for the young scientist," one of the old guard declared not long ago.

Sydney Brenner qualified that: "I hold the somewhat weaker view that history does exist for the young, but is divided into two epochs: the past two years, and everything that went before."

E. O. Wilson believes that this short-term memory may be a good thing. He contrasts it with the veneration that psychologists pay to Freud and Jung or that social theorists still pay to the heroes in their pantheons. "Much of what passes for social theory is still in thrall to the original grand masters," Wilson wrote recently, "—a bad sign, given the principle that progress in a scientific discipline can be measured by how quickly its founders are forgotten."

As Jeff Hall approaches the climax of his story, leading the ranks up the path of Pickett's Charge, a few molecular biologists, absorbing the spirit of the place, are thinking and talking about the beginnings of their own charge. The Civil War was the moment in time during which Mendel's peas and T. H. Morgan himself were, as Morgan used to say, "laid down." The chairman of Princeton's molecular biology department this year, Arnie Levine (they call him General Levine today), reminisces with Lee Silver about Schrödinger's *What Is Life?* Schrödinger speculated in his book that quantum jumps might cause mutations. "It was wrong," Levine says with a laugh. "But that didn't matter. It got the physicists in."

"Brought in Francis Crick," Silver agrees. "Seymour Benzer . . . Gunther Stent. . . ."

But like most molecular biologists, Silver prefers looking into the future. He feels his science is racing toward a climax. "We thought there were all these barriers, and they don't exist," he often says. "We're finding things we thought we'd never be able to find. Barriers to knowledge keep disappearing one by one. I think in the end we're going to know it all. I really do. It's just a question of how long, just a question of when."

Soon it will be straightforward to take a small sample of someone's DNA and use an electronic device called a DNA chip to probe for variant forms of every single gene. At a glance a molecular geneticist will know what genes that individual carries and what genes are on right now and what genes are off. Genetics start-up companies are already manufacturing the first generations of these DNA chips in a union of computer science and molecular biology that seems likely to race ahead at the customary speed of both.

By screening the DNA of one hundred thousand people, combining that information with personality tests, and letting a computer crunch it all together, molecular geneticists will put together pictures of gene complexes working together to produce the most complex traits of personality. "It's going to happen fast!" Silver says. "If there are ten genes that make somebody aggressive, you're going to *see them!*" The twentieth century began with a man looking at one white-eyed fly in a bottle, he says. Before we are that far into the twenty-first, he says, "we'll be able to take ten thousand people and match different combinations of alleles across the whole genome and come up with a behavioral profile." Of course, each profile will have been modified by the environment. "That's certainly the case," Silver says. "But it's an incredible story. I think people don't realize the power of genetics. You can figure out which genes are responsible for a trait—without knowing anything." Knowing nothing about the gene, the environment, the psychology, or the physiological machinery, you can find your way in. "Knowing *nothing*! Because once you figure out what the connection is, then you go back and figure out *why*. You can do all that afterwards." Take shyness, which Silver believes is very much genetically determined. With the kind of mass screen that Silver is envisioning, he could find two dozen genes, each with multiple alleles, that contribute to shyness. He could do that without knowing what each gene does. "*Then* you can ask, What does it do? What protein does it make? When is it turned on? When is it turned off? Incredibly powerful, to do all this.

"People who don't believe in relativity don't *understand* relativity. People who don't believe in evolution don't *understand* evolution. And it's the same with genetics. And I think some people are just reluctant to let their imaginations run.

"My feeling is that molecular biologists are going to move into psychology and take over the field. I think that's the way psychology is going to be rejuvenated.

"In the 1970s, they said genetic engineering would be impossible. Then they said cloning would be impossible. Amazing that people can be so shortsighted. It's an *explosion* of science. Right now we're really at the beginning of biology. That's really the way to look at it. The end of biology in this century is like the end of physics in the last century."

JEFF HALL holds his megaphone tipped at a rakish angle to fire out the sound over the heads of the front rows of the crowd. He is bulling into the microphone, angrily miming the action. He is almost at the top of the path of Pickett's Charge. The statue of General Lee is far behind him, and the statue of George Meade is just ahead, beyond the Clump of Trees. In Hall's view, the Civil War was the greatest drama in the history of the United States; the battle at Gettysburg was the climax of the war; and "the climax of the climax, the central moment of our history," as one of Gettysburg's historians puts it, was Pickett's Charge. On July 3, 1863, fourteen thousand Confederate soldiers marched up this slope, toward what is now known as Cemetery Ridge. They marched through cannon fire and rifle fire toward a bend in a low stone wall at the top, now known as the Angle. They advanced, flags waving, into the very center of the Union lines at the top of Cemetery Ridge. Only two hundred of Pickett's troops made it to the Angle, about the size of the band Hall is leading up the path today, counting the children. In that hour the battle was lost and won.

IT IS ALREADY POSSIBLE—in fertility clinics it is done every day—to screen the DNA of a set of eight embryos at the eight-cell stage and let the parents pick the one they want to implant in the mother's womb. The more genes there are to screen and the better these gene complexes are understood, the more wealthy parents will select not only the healthiest but also the best and the brightest embryo they can, designing the genes of their children. With the same tools that Hall used to inject the first instinct into an animal, it may someday be possible for people in fertility clinics to inject a wide selection of human instincts and traits as well. As these choices are made more and more often, the old dream of Galton and the eugenicists who followed him will be fulfilled willy-nilly over the next few centuries whether governments legislate for it or against it. The rich will pick and choose the genes of their children; the poor will not. The gap between rich and poor may widen so far in the third millennium that before the end of it there will be not only two classes of human beings but two species, or a whole Galapagos of different human species. These human species could be prevented from interbreeding by the genetic engineering of chemical incompatibility, so that the egg of one would reject the sperm of the other. Silver is thrilled by the power of his science and by the vision of barriers falling away, and yet looking into the far future he sometimes thinks he sees disaster, a Darwinian nightmare; out of utopian eugenics, a dystopian origin of species.

"We have reached this point down a long road of travail and self-deception," E. O. Wilson wrote recently. "Soon we must look deep within ourselves and decide what we wish to become. Our childhood having ended, we will hear the true voice of Mephistopheles." He is sure we will not want to turn ourselves into protein-based computers; we will not want to lose what makes us human. Wilson's ants, for instance, never play. We will not want to give up what we have evolved over billions of years, going back to the very origin of life. But what changes will we make in our natures—deliberately or casually and without plan—beginning in the next few years? "What lifts this question above mere futurism," Wilson writes, "is that it reveals so clearly our ignorance of the meaning of human existence in the first place."

ON·MOST NIGHTS, by nine o'clock, Benzer is almost alone on his floor of Church Hall. By ten or eleven, his desk lamp is one of the last lights burning in the windows. For Benzer the smell of the fly food has never lost its savor: home, sweet home.

Sometimes when he is alone in his lab, he thinks of putting up plaques on the doors with the names of the people who worked there. They made his revolutions their beacon, and some of them found harbors and some lost their ships. Not all of Max Gottlieb's Arrowsmiths stay in the game—and some of Benzer's first were the first to drop out. Konopka now lives a few blocks away from the Caltech campus, alone in a small house half hidden by palm trees and magnolias, as anonymous as Kafka's K. Once in a while he gets a new clock paper in the mail. He looks at the tables and he thinks, "Well, heck, these are all my mutants!" He spends his days collecting butterflies now, tipping his forehead sharply forward to peer at them over his glasses. He also collects Grateful Dead tapes and photographs of local waterfalls. He has a big Lionel model train set on the shag carpet by his front door and a pinball machine shoved against the dining room wall: "Gottlieb's FAR OUT." Back in the summer of 1968, the scientists in Church Hall said Konopka would never find what he was looking for. Then, when he found it, they said it was meaningless. Now it is meaningful, and he is out of science. "Story of my life," he says. "They just didn't believe it. You think scientists are open-minded, but ha, ha, ha."

In the middle of the night in Church Hall, Benzer often wanders up to T. H. Morgan's filing cabinets in the third-floor hall to raid and riffle through the founding papers of genetics, ancient references he wants to see. Then he stops in to see Sturtevant's old student Ed Lewis. Staring at the baby octopi in Lewis's aquarium tank at two in the morning, Benzer is filled with the feelings that sometimes come to him down at Sturtevant's bed of experimental irises. The same thought possesses him even standing over a puddle, thinking of all the microscopic vorticelli and rotifers and pond creatures doing their tricks in there. "It's a wonderful, fabulous world, and it's been kicking around a long time," he says. "And there's so much going on all the time. It's just amazing how much we're neglecting."

Through his microscope Benzer zooms in on the eye of the fly. He admires one facet. On this facet there is a hair, and the hair has a nerve that goes into the brain. If he looks at it closely enough, the single facet starts to look like the whole eye. Even through an electron microscope at 20,500 times magnification, he still sees more fine structure. "The more you look, the more you find," he says. And this is just looking at the surface. Looking inside, you find worlds within worlds of detail: coils and coils of wires, cables, and corrugated gooseneck tubing, buttons and corks, tufts, four-leaf clovers, and odd projections like golf balls on tees. "The eye is the microcosm that contains all of biology in it. Maybe even including consciousness," says Benzer. "But that's the way it is; every kernel contains all of biology, practically." Feynman once said it beautifully: "Nature uses only the longest threads to weave her patterns, so each small piece of her fabric reveals the organization of the entire tapestry."

> *Behavior was fun, but I don't care,*
> *I'm on to something else next year,*
> *I must stick with the new frontier*
> *Until I'm old and gray.*

Now here he is old and gray, and once again behavior is looking to him like the new frontier. So is aging, the whole phenomenon of lifespan: How much of that is in the genes? Once again he is wondering if science can find answers to the recurrent questions that Max despaired of answering—find something worth telling Aurora, the goddess of the dawn.

By 3 a.m., the Church Laboratory is very quiet. At the far end of the hall and down one flight of stairs, a small night shift of lab technicians is washing flasks, racking test tubes, and filling fly bottles for the next day's rounds of experiments, including hundreds of vials, test tubes, and antique milk bottles marked BENZER. At four o'clock in Church Hall, the night shift will go home. At six o'clock, the day shift will arrive, in the form of a woman whose name happens to be Aurora. And Benzer, going out the door, will meet Aurora, coming in.

Not long after Benzer married Carol Miller, Francis Crick asked her to show him the human brain. Crick had been thinking and theorizing

Just after Benzer's seventy-seventh birthday, he and two of his postdocs, Yi-Jyun Lin and Laurent Seroude, announced the discovery of a mutant fly that lives more than one hundred days. They named it methusaleh. Other drosophilists had shown that flies' lifespans are influenced by genes, but none of them had ever cloned one. Now Benzer plans to hunt for more lifespan mutants. He is embarking on the new career he hopes to pursue in his eighties: the genetic dissection of aging. When he introduces methusaleh in lectures, Benzer shows a slide of a mutant with a fly's eyes and Darwin's beard.

about the brain for a few years, but he had never actually seen one. He wanted Carol to show him the cerebral cortex so that he could see the edges.

So Carol set a brain on an ordinary white plastic cutting board from the hardware store, while Crick watched in a borrowed white lab coat. Around them the cabinets and shelves held dozens of brains, some intact, some already dissected to bits. On the shelf above the cutting board was a row of straight-sided jars full of grayish eyeballs.

The convolutions of the cerebral cortex are like the coiling and super-coiling of the double helix: origami tricks by which evolution has managed to pack a great deal of information into a very small space. Carol explained that if she could spread it out, its edges would fall off the cutting board on all sides—almost a square yard of cortex. There were millions of nerve cells packed into every square inch, each nerve making thousands of contacts with its near and distant neighbors according to patterns that were laid down first by the genes and the growing nerves in the embryo, then by a lifetime of choices inside that gray-brown sheet.

Quite a contrast with the brain of the fly, which fits into a head case so small that it is hard for us to see without a magnifying glass.

The reasons that we evolved such massive brains remain obscure, but one reason may have been to help us at the choice points. Our brains allow each of us to bring a maximum amount of learning and experience to each and every choice point, all that our species has learned and all that we have learned in our lifetimes. A fly does this to some small degree, and we do it to a large degree—more than any other creature on this planet.

For some years now, Crick and a few colleagues have been trying to figure out the difference between unconscious vision and visual awareness. Crick assumes that there must be some difference in the way incoming information is processed that determines whether we are aware of it or whether we are not. By tracing this difference in the brain, Crick hopes to find a clue to the way in which any experience can become conscious, allowing us to get maximum benefit from our big brains at the choice points. Crick thinks this will be the problem of the twenty-first century, although Benzer raises his eyebrows and smiles his molecular smile whenever they talk about it.

"He teases me because I'm interested in consciousness and so on," Crick says with a laugh. "And of course in the case of *Drosophila* it wouldn't be very sensible, because we've no idea—we hardly know what it means to be conscious in a *mammal*; when we get down to *Drosophila*,

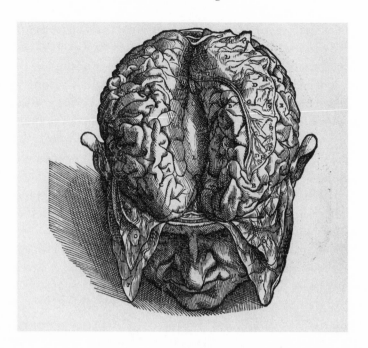

*Cerebral cortex, drawn by the anatomist Andreas Vesalius. Vesalius published his* De Humani Corporis Fabrica *in 1543, the same year that Copernicus published his* De Revolutionibus Orbium Coelestium. *Copernicus opened a journey outward, Vesalius a journey inward. The journey outward has now led to the discovery of light from the Big Bang and the birth of the universe. The journey inward has led to the discovery of the first links between genes and behavior. Someday these discoveries too will be remembered as beginnings, first openings, points of departure.*

we really don't know whether they're automata or not. So I can see why he doesn't feel really interested himself in that topic. And I wouldn't be if *I* worked on *Drosophila*." "You know," Crick once joked from a podium at a meeting in Pasadena, "Jacques Monod used to say that everything that was true of *E. coli* was true of the elephant. But I don't think that even *he* said that everything that was true of the elephant was true of *E. coli*. I don't necessarily think the fly is as smart as Seymour, even though Seymour doesn't know how to land on the ceiling."

"If you will be good enough as to give me a definition of consciousness," Benzer retorted (from the floor), "then I will try to devise a test to see whether it is present in *Drosophila*. But so far you have been unable to come up with a definition."

Crick hopes that studies of the human brain's visual processing system will lead him there. He assumes that there is also a part of the brain devoted to planning and looking ahead, and here he suspects the frontal lobes. The frontal lobes are the most forward part of the human brain, the part just behind the forehead. The very frontmost portion of the frontal lobes, the prefrontal fibers, are thought to be sites of our social reins and bridles; they keep us from saying and doing things that veer off the path of what is socially expected and accepted. The archetypal case of frontal lobe dementia was one Phineas Gage, foreman of a work crew on the Rutland and Burlington Railroad in Cavendish, Vermont. Gage lost much of his frontal lobes in a dynamiting accident on September 13, 1848, when an iron bar rammed into his head just below the left eye, shot through his skull, and flew out through the crown of his forehead, smeared with blood and brains. He astonished his crew by walking, talking, and joking soon after the accident, although it soon became clear, as his doctor's notes show, that he was "no longer Gage":

> October 15 (32nd day) . . . Intellectual manifestations feeble, being exceedingly capricious and childish, but with a will as indomitable as ever; is particularly obstinate; will not yield to restraint when it conflicts with his desires.

> October 20th (37th day) . . . Sensorial powers improving and mind somewhat clearer, but very childish.

> November 15th (64th day) . . . Is impatient of restraint and could not be controlled by his friends.

Studies of patients with damage caused by strokes have suggested that frontal lobes have very specific and localized functions. A lesion in one place produces disinhibition like Gage's, but a lesion in another place produces apathy. (The prefrontal lobotomy works because it induces apathy.) A lesion in a third place produces blindsight. Subjects with blindsight can see, but they claim not to. They can point to objects around the room when an experimenter asks them to, but they will deny that they can see them. Their eyes work and their brains can process the information, but they are no longer conscious of the results. Carol demonstrates these areas of the brain on a cutting board with the same mix of expressions, half matter-of-fact, half reverent, that Benzer has when he gives guided tours of the brain of a fly. "This is the frontal cortex," she says. "This is the dura mater. Here is part of the basal ganglia.

The apathy area is here. And disinhibition is down here. Frontal tempo-ral dementia."

She has begun to apply Seymour's method of genetic dissection to the frontal lobes. There are forms of frontal lobe dementia that run in fami-lies; one form of the disease with late onset (in the fifties or sixties) seems to be linked to chromosome 17. By dissecting the brains of victims of the disease in autopsies and staining them with fly stains to examine fine-scale changes under the microscope, she is trying to trace the links from gene to behavior just as Seymour does in flies. Through the microscope her brain sections stained with fly stains are abstract landscapes, some of them like stylized artist's impressions of a tree stained pink and some rather beautiful aerial landscape views, with alluvial curves and scallop-edged patterns. The stain swivels and follows the contours and the curves, oblivious to the pathos of mortality. Somewhere in there, Crick believes, may be the answer to the problem of free will.

THE ROMAN PHILOSOPHER Lucretius imagined that atoms must swerve somehow; "if the atoms never swerve so as to originate some new move-ment that will snap the bonds of fate, the everlasting sequence of cause and effect—what is the source of the free will possessed by living things throughout the earth?"

At the end of the day in Cold Spring Harbor, Watson also thinks in terms of some kind of Lucretian swerves. "My hypothesis is that free will comes from the imperfect working of the brain," he says. "The machine is *inherently* uncertain." He smiles at the pun.

"But on certain occasions I *know*," he says, alluding here to a particu-larly trying meeting he has just endured, a board meeting in which he has listened to the president of a new biotechnology company that Watson thinks is mismanaged. "You know, when I'm in a room, and I hear shit, after a while the word 'shit' is going to come out. You just can't take it any-more. Now that's, hmmm, a predictable response. It's *bound to come out.* I think to myself, maybe I'll sit through nonsense and not say it. But . . ." Watson sighs. "So in that sense you don't have a free will. Your reactions are programmed. You know, you start asking the difference," he says with a nod toward the Fly Room next door. "What free will is there in *Drosophila?* You put the question of the free will of a *fly.* And what's really different about the fly's brain from ours—which gives us free will?

"I'm sure once we know how the brain works, we'll no longer talk about free will in the Jesuit sense. It will cease to be, you know—" Freedom will cease to be a mystery requiring Jesuitical debate; it will cease to be a theological or philosophical question. "It will just be how the brain works. You will describe how the brain works. You won't use the words 'free will'; you know, you'll *understand*. . . . Because you're asking, how does the brain work?" he says in a softer voice. "That's what you're *really* asking." The bell in the double-helical bell tower outside his office window begins to toll. "And that's really the ultimate question to ask," he says, speaking through the tolling of the bell with a pleased chuckle in his voice, suddenly sounding very much like his old friend Crick.

WHEN BENZER himself thinks about the free-will question, he always remembers his first moments watching the flies in his test tubes run toward the light. In the very first experiment he did, most of the flies went to the light, but some didn't. He tested them again; most went to the light, but some didn't. That was why he made the countercurrent apparatus. "If you mean a certain randomness in behavior—then flies have free will," Seymour says sometimes at the Red Door Cafe when the talk turns to drosophilosophy. Why did each of those individual flies make each of those individual decisions and revisions? "That's free will if you want to call it that."

But when he talks this way at the Red Door, his postdocs are likely to give him the same raised eyebrows and molecular smiles they have learned from Benzer. "If flies had free will," they tell Benzer, "your lab would be empty."

Benzer gets no further when he talks about this question with Carol, describing the zigzags of his flies in the countercurrent machine. "So what is that?" he asks. "Is that free will? If the fly makes up its own mind?"

"Well, the problem is, you don't know what its mind is to make up," Carol says.

"I'd say free will means, if you are subjected to identical stimuli, you don't necessarily do the same thing . . ."

And so they go around and around, even people who live with their hands in rubber gloves deep in genes and brains. Perhaps it is still too soon to talk about the question. A sense of paradox still hovers over the whole picture of life, from the smallest scale to the largest.

How is it possible that these two things should be true at once? Schrödinger asks in *What Is Life?*: first, that his body is a mechanism, a clockwork, and runs according to the laws of nature; second, that he knows "by incontrovertible direct experience" that he is running the show, that he is directing the motions of his body, that he can foresee what he is doing, and that he can take responsibility for his actions. As he struggles to answer these questions in the last pages of *What Is Life?*, Schrödinger begins to sound a little like Berger on the wall of Seymour's Sandwich Shop. He quotes Schopenhauer and Kant; the cultural milieu (*Kulturkreis*); the Upanishads' recognition that ATHMAN = BRAHMAN (the eternal self); and the Christian mystics' phrase DEUS FACTUS SUM (I have become God). Schrödinger concludes that there is only one consciousness "and that what seems to be a plurality is merely a series of different aspects of this one thing, produced by a deception (the Indian MAJA); the same illusion is produced in a gallery of mirrors, and in the same way Gaurisankar and Mt. Everest turned out to be the same peak seen from different valleys."

Each of us feels like an "I." "*What is this 'I'?*" Schrödinger asks. How is it that each of us feels like one single person even though we began life so long ago that we have lived through a succession of more identities than the instars of insects, and the world of our first memories feels to us like a distant country? In spite of all the living and all the forgetting that we have done, and in spite of all the mechanisms that science has discovered over our heads and inside our heads, we are all still here living life. "In no case," Schrödinger wrote, "is there loss of personal existence to deplore."

And, he concluded, "Nor will there ever be."

NERVES, as they grow in an embryo, seem to wander like flies in a countercurrent machine. Each of them is guided by genes, yet on each scale there seems to be some play in the system.

Likewise, there seems to be play in the system for each of us as we make the choices we do. "A human behavior pattern is not a monument to a life that is gone, but a drama full of life," writes the philosopher Abraham Heschel. "It is a system as well as a groping, a wavering, a striking forth; solidity as well as outburst, deviation, inconsistency; not a final order but a process, conditioned, manipulated, questioned, challenged, and guided."

"Diversity is as wide as all the tones of voice, ways of walking, coughing, blowing one's nose, sneezing," writes Pascal. "We first distinguish grapes from among fruits, then muscat grapes, then those from Condrieu, then from Desargues, then the particular graft. Is that all? Has a vine ever produced two bunches alike, and has any bunch produced two grapes alike?

"I have never judged anything in exactly the same way," he continues. "I cannot judge a work while doing it. I must do as painters do and stand back, but not too far. How far then? Guess. . . ."

Even in the behavior of our thoughts, there is play in the system. We all feel them zigzag, heading for the light, but on again, off again, like the fly in the countercurrent machine. "Thoughts come at random, and go at random," Pascal wrote. "No device for holding on to them or for having them.

"A thought has escaped: I was trying to write it down: instead I write that it has escaped me."

Emerson wrote, "Thoughts come into our minds by avenues which we never left open, and thoughts go out of our minds through avenues which we never voluntarily opened."

"I am lying in bed, for example, and think it is time to get up," wrote William James,

> but alongside of this thought there is present to my mind a realization of the extreme coldness of the morning and the pleasantness of the warm bed. In such a situation the motor consequences of the first idea are blocked; and I may remain for half an hour or more with the two ideas oscillating before me in a kind of deadlock, which is what we call the state of hesitation or deliberation.

Somehow or other, after lolling around in the bed like this for half an hour, the philosopher confessed, "I shall suddenly find that I have got up." It is as if his mind had a mind of its own. Is his springing out of bed free will? ("Free will," Seymour says to Carol, "is if you get back in.")

Even with the wanderings of science, there is play in the system. Science blunders along, like every sort of behavior. Max Delbrück knew that science is always improvisatory: "The grand edifice of Science, built through the centuries by the efforts of many people in many nations, gives you the illusion of an immense cathedral, erected in an orderly fashion according to some master plan. However, there never was a master plan. The edifice is a result of channeling our intellectual obsessive forces into the joint program. In spite of this channeling, the progress of science at all times has been and still is immensely disorderly for the very reason that there can be no master plan." In science, as in the rest of life, the paths are paths only in retrospect.

And in the tree of life itself there seems to be play in the system: what look like swerves and random branches. The shape of a growing nerve tree in the brain, the shape of the decision trees of an individual lifetime, and the shape of the whole tree of life share the same branching form on successively larger and larger scales. What is the Tree of Life but a decision tree, a drawing of a series of choice points in which one line of life went one way and another another way? And were those choices forced or free? Some of them were based on new behavior at choice points, as when a fish first began hanging around longer and longer at the margins of the sea and taking gulps of air.

"The things we thought would happen do not happen;/The unexpected God makes possible," Euripides wrote in the last lines of one of his last plays. Since we see paradox at all scales, is it possible that the answer to this paradox on one scale will someday unlock them all? In his laboratory, Jeff Hall is trying to move from the particular to the general, exploring more and more interconnections among the time, love, and memory genes, continually e-mailing Tim Tully and Ralph Greenspan, who was Hall's first graduate student, now at the Neurosciences Institute in San Diego. One of their clock genes seems to mesh with one of their memory genes like the teeth of two gears. Clock genes also seem to mesh with the genes that build the body of the fly from the egg. And the human *period* gene is expressed not only in the human brain but in the pancreas, kidney, skeletal muscle, liver, lung, and placenta—almost everywhere its discoverers have looked for it. Although no one is sure what that means, they are sure that the fly clock is interwoven with the rest of the fly's genes and that the human clock is interwoven with much of what is human.

In Tokyo, Yoshiki Hotta, who helped Konopka find the clock mutants, is also thinking about the way the genes fit together. So is another old student, Alberto Ferrús, in Madrid. Ferrús thinks we stand in relation to the atomic theory of behavior now the way biologists stood in relation to the atomic theory of inheritance at the turn of the twentieth century. Then, biologists had accumulated an enormous amount of information about what happens when you cross different species of plants or different species of animals. But they did not understand the basis of heredity. Then, in 1900, they rediscovered Mendel's laws. When these laws were rediscovered, vast amounts of information were wiped out because there was no need for them anymore. Biologists now had an organizing principle by which they could understand heredity in any organism. "And maybe we need for neurobiology a similar type of breakthrough," says

Ferrús. "We would like to find some sort of neural code to understand how perception is encoded in our brains, for instance." Or how we memorize a piece of information we have perceived. "Code. How does a brain of a fly or a man operate? The basic principle. The details are going to be different in every organism, of course. But maybe, maybe there is such a basic principle, such a code that it is of universal value. And if such a thing is once discovered, then we would have made a gigantic step ahead, as gigantic as Mendel's laws were for genetics."

The next generations of genetic dissectors are beginning to think about systems of genes. They are beginning to think about genes the way Konrad Lorenz thought about instincts: "Unless one understands the elements of a complete system as a whole, one cannot understand them at all." Like the rest of molecular biology, genetic dissection has always pushed toward the particular. The genetic scalpel uncovered one gene in one behavior at a time. But each gene is involved in many kinds of behavior. Hall, Tully, and Greenspan are writing a book on the subject of genes as networks, genes as systems, genes as constellations. They mail their manuscripts back and forth, struggling toward a difficult vision that they cannot put into words. One day Greenspan scribbled in his draft in huge letters: "I'M GROPING, NOT TRACKING HERE. There's an idea struggling to get out. HELP!!!" Sometimes Greenspan seems to get a dim glimpse of a web of interweavings, a web that feels something like a new view of life and something like inner experience. Being alive does not feel like an assortment of conflicting separate instincts, pieces, and inclinations; we feel, when we introspect, that there are many parts but that they are at least loosely woven together, loosely connected in what Lorenz called "the great parliament of instincts." In life, emotional wisdom resides in seeing around particles of mood, moments of mood, to the whole in time, putting it all together. In life we are not particles but evolution. The word "religion" comes from the Latin *religare,* "to bind loose things together." Hall, Greenspan, and Tully struggle with thoughts of networks; the book will not come together. And Hall escapes more and more to Gettysburg and drives home with the bumper sticker SAVE THE BATTLEFIELDS.

Could it be that the answer to the paradox has something to do with the way we break the world into categories, and will we see the answers to all these paradoxes when we have learned to think in new categories? Maybe the free-will problem has to do with categories of thought we cannot see because they are innate in us, because they are part of the instincts with which we are born.

"What is your aim in philosophy?" Ludwig Wittgenstein asked in his *Philosophical Investigations* in 1953, the year that Watson and Crick put together their model of the double helix, and Benzer figured out how to map the interior of a gene. The philosopher answered himself, "To show the fly the way out of the fly bottle." Well, the fly is out of the bottle now. Whether we like it or not, the fly zigzags through all our meditations "like crack through cup," as the poet Rilke writes in one of his *Duino Elegies*. In ways we may love or hate, the science that came out of the fly bottle is changing our sense of what it means to live. And maybe the science will help us understand this overarching paradox. This is the question that the fly bottle has posed from the beginning, and maybe the answer, too, will come out of a fly bottle. "Who's turned us around like this?" Rilke asks in his elegy. "Who's turned us around like this, so that we always, do what we may, retain the attitude/of someone who's departing?" Maybe the fly will lead, in someone's night thoughts, to a new union of Pascal's two infinites. Maybe the fly will lead us out of the bottle into territory that is as blurred and vague now as the gene was before the first flies flew in the window.

Butler's line about eggs making eggs is a variation on the theme of the old riddle: Which came first, the chicken or the egg? Maybe the answer to that riddle will prove to be the answer to them all: some kind of engagement, collaboration, development involving both gene and world. And maybe it is the same with consciousness. Maybe all these branches have the same shape because they are all products of some interaction, some drama or dance of life and world that we have yet to find a way to wrap our minds around. Maybe this is the way out of the fly bottle of the twentieth century. Our genes and brains work only by engagement with the world around them, so we are not imprisoned by our genes and brains. "Denmark's a prison," says Prince Hamlet. "Then is the whole world one," his college friend replies. If our genome is a prison, then is the whole world one.

This is the kind of problem that more and more neurobiologists and neurophilosophers are trying to solve, but Benzer only raises his eyebrows and smiles his molecular smile: "Oh, they can have it. I'll leave it to them."

# Notes

Between 1994 and 1998, I spoke with almost 150 biologists who have made, or watched the making of, what I call in this book the atomic theory of behavior. With some of them I spoke once, with some many times. Unless otherwise noted, all quotations in the book come from those interviews.

PAGE

vii **"The Fly"**: William Blake, *The Portable Blake* (New York: The Viking Press, 1946), 107.

### PART ONE: OCCAM'S CASTLE

3 **"Would it be too bold"**: Erasmus Darwin, *Zoonomia; Or the Laws of Organic Life*, 4th American ed. (Philadelphia, 1818), I, 397. Quoted in John C. Greene, *The Death of Adam* (Ames, Iowa: The Iowa State University Press, 1959), 167.
  **"My handwriting"**: Charles Darwin, *Metaphysics, Materialism, and the Evolution of Mind: Early Writings of Charles Darwin*, transcribed and annotated by Paul H. Barrett, with a commentary by Howard E. Gruber (Chicago: University of Chicago Press, 1980), 21.

### CHAPTER ONE: FROM SO SIMPLE A BEGINNING

5 **"The ancient precept"**: Ralph Waldo Emerson, "The American Scholar," in *Essays and Poems* (New York: The Library of America, 1996), 56.
  **"The nearest gnat"**: Walt Whitman, "Song of Myself," *Complete Poetry and Collected Prose* (New York: The Library of America, 1982), 84.
6 **"legs with joints"**: Blaise Pascal, *Pensées*, A.J. Krailsheimer, trans. (London: Penguin Books, 1995), 60.
  **"The eternal silence"**: Ibid., 66
7 **"The way in which"**: Ibid., 65. The passage is from Saint Augustine, *City of God*, XXI.10
  **like panpipes**: For his own description of this experiment, see Benzer, "Behavioral Mutants of *Drosophila* Isolated by Countercurrent Distribu-

tion," *Proceedings of the National Academy of Sciences USA* 58 (1967): 1112–19. The paper was reprinted in 1994 with interviews and commentaries in the "Landmarks" series of *The Journal of NIH Research* 6: 66–73. Very little has been written about this work outside the technical literature. Benzer has published one popular article about his research: "Genetic Dissection of Behavior," *Scientific American* (December 1973): 24–37. For a witty and accessible historical review of the development of the field, see Ralph J. Greenspan, "The Emergence of Neurogenetics," *Seminars in the Neurosciences* 2 (1990): 145–57. See also R. J. Greenspan, "Understanding the Genetic Construction of Behavior," *Scientific American* 272, no. 4 (1995): 72–8.

7  **Benzer got the idea:** My account is based on several extended visits and interviews with Benzer in Pasadena between 1995 and 1998. I have also drawn on transcripts of an excellent series of unpublished interviews that Heidi Aspaturian of Caltech conducted with Benzer as part of an in-house oral history project in 1990 and 1991.

8  **Aristotle mistook them:** For two vignettes of early fly history, see B. Peyer, "An Early Description of *Drosophila*," *Journal of Heredity* 38 (1947): 194–9; and G. H. Müller, "*Drosophila*: A Contribution to Its Morphology and Development by W. F. von Gleichen in 1764," *Journal of Natural History* 10 (1976): 581–97. Aristotle describes the flies in *Historia Animalium*, Book 5, Chapter 19 (cited in Peyer).

   **first published laboratory study:** F. W. Carpenter, "The Reactions of the Pomace Fly (*Drosophila ampelophila Loew*) to Light, Gravity, and Mechanical Stimulation," *Contributions from the Zoölogical Laboratory of the Museum of Comparative Zoölogy at Harvard College* 162 (1905): 157–71. For a discussion of the experiment's place in the early history of fly work, see Garland E. Allen, "The Introduction of *Drosophila* into the Study of Heredity and Evolution: 1900–1910," *Isis* 66 (1975): 322–33.

9  **happy medium:** S. Benzer, "From the Gene to Behavior," *Journal of the American Medical Association* 218 (1971): 1015–22. Presented as an Albert Lasker Basic Medical Research Award Lecture, New York, November 11, 1971. The award was for his work on splitting the gene, but Benzer took the occasion to talk about his new passion, the study of behavior.

   **about one hundred thousand neurons:** Ibid. No one has ever actually counted. This number is an order-of-magnitude estimate; the actual number may be three times higher.

10  **"There is enough light":** Pascal, *Pensées*, 50.

12  **"from Crudity to Perfect Concoction":** Francis Bacon, *Sylva Sylvorum; Or a Naturall Historie* (1626), 838; quoted in *The Compact Edition of the Oxford English Dictionary* (Oxford: Oxford University Press, 1971), 3459.

   **"that sweetness is not bitterness":** John Locke, *An Essay Concerning Human Understanding* (Amherst, N.Y.: Prometheus Books, 1995 [1690]), 19.

13  **"Have you not noticed":** Freud, "The Question of Lay-Analysis," first published in 1926, translated in Volume 20 of *The Standard Edition of the Complete Psychological Works of Sigmund Freud*, J. Strachey, ed. (London: Hogarth Press, 1953–1974); excerpted in *The Study of Human Nature*, Leslie Stevenson, ed. (Oxford: Oxford University Press, 1981), 164–192.

**One of Tinbergen's books:** Niko Tinbergen, *The Study of Instinct* (Oxford: Oxford University Press, 1969 [1951]), 78. See also N. Tinbergen, "Social Releasers and the Experimental Method Required for Their Study." *Wilson Bulletin* 60 (1948): 6–52.

**a new synthesis:** For a time-capsule overview of the controversy, see Arthur L. Caplan, ed., *The Sociobiology Debate: Readings on Ethical and Scientific Issues* (New York: Harper & Row, 1978). For a glimpse of the controversy today, see S. Pinker and S. J. Gould, "Evolutionary Psychology: An Exchange," *The New York Review of Books* (October 9, 1997): 55–7.

14 **"There is grandeur":** Charles Darwin, *On the Origin of Species* (London, 1859; Facsimile ed. Cambridge, Mass.: Harvard University Press, 1964), 490.

15 **"To compare the speed":** John Maddox, "Valediction from an Old Hand," *Nature* 378 (1995): 521–3.

CHAPTER TWO: THE WHITE-EYED FLY

17 **"Those who love wisdom":** Heraclitus, Frag. 3. Quoted in J. T. Fraser, *Of Time, Passion, and Knowledge: Reflections on the Strategy of Existence* (New York: George Braziller, 1975), 3.

**"but what is really happening":** S. Benzer, personal communication.

**He found Democritus:** Robert Burton tells this story in "Democritus Junior to the Reader," the preface to his *Anatomy of Melancholy* (Michigan State University Press, 1965 [1621]), 8. Democritus's heroic but doomed effort inspired Burton to write his book, he says: "to revive again, prosecute, and finish in this treatise."

18 **delivered a young goat:** Quoted in Robert J. Richards, *Darwin and the Emergence of Evolutionary Theories of Mind and Behavior* (Chicago: The University of Chicago Press, 1987), 21.

**Some of the symbols:** Carl G. Liungman, *Dictionary of Symbols* (Santa Barbara, Calif.: ABC-CLIO, 1991).

**"What a wonderful thing":** Quoted in François Jacob, *The Logic of Life: A History of Heredity* (Princeton, N.J.: Princeton University Press, 1993), 20.

19 **"A devil, a born devil":** W. Shakespeare, *The Tempest,* Act IV, Scene 1, in *The Comedies and Tragedies of William Shakespeare,* vol. 1 (New York: Random House, 1944), 51.

**"instance of hereditary mind":** Darwin, *Metaphysics,* 21.

**Galton later closed his copy:** Francis Galton, *Memories of My Life,* 3d ed. (London: Methuen & Co., 1909 [1908]), 288.

**Gregor Mendel:** Vítezslav Orel, *Gregor Mendel: The First Geneticist,* Stephen Finn, trans. (Oxford: Oxford University Press, 1996).

20 **climbed Mount Fuji:** Aspaturian interviews, 143–4.

**"A gentleman of considerable position":** Quoted in Charles Darwin, *The Expression of the Emotions in Man and Animals* (Chicago: University of Chicago Press, 1965 [1872]), 34.

21 **"We seem to inherit":** Francis Galton, *Natural Inheritance* (London: Macmillan and Co., 1889), 7.

**Mendel's paper might:** Among many accounts of the birth of genetics, I relied on A. H. Sturtevant, *A History of Genetics* (New York: Harper &

Row, 1965). For an account that sets these events in a wide historical context, see Ernst Mayr, *The Growth of Biological Thought* (Cambridge,
Mass.: Belknap Press, 1982). For a clear and colorful historical overview,
see John A. Moore, "Science as a Way of Knowing—Genetics," *American
Zoologist* 26 (1986): 583–747.

21 **In the fall of 1907:** The literature on Morgan's Fly Room is rich. My chief
sources were Sturtevant, *History of Genetics*; Garland E. Allen, *Thomas
Hunt Morgan* (Princeton, N.J.: Princeton University Press, 1978); Robert
E. Kohler, *Lords of the Fly: Drosophila Genetics and the Experimental Life*
(Chicago: University of Chicago Press, 1994); Elof Axel Carlson, *The
Gene: A Critical History* (Philadelphia: W. B. Saunders Company, 1966);
E. A. Carlson, "The *Drosophila* Group: The Transition from the
Mendelian Unit to the Individual Gene," *Journal of the History of Biology*
7, no. 1 (1974): 31–48; and Nils Roll-Hansen, "Drosophila Genetics: A
Reductionist Research Program," *Journal of the History of Biology* 11, no. 1
(1978): 159–210.

**breed animals in the dark:** Fernandus Payne, "Forty-nine Generations in
the Dark," *Biological Bulletin* 18 (1910): 188–90.

22 **tin cans:** Curt Stern, "The Continuity of Genetics," *Daedalus* 99, no. 4
(1970): 882–907.

**student cafeteria:** Ian Shine and Sylvia Wrobel, *Thomas Hunt Morgan:
Pioneer of Genetics* (Lexington, Ky.: University Press of Kentucky, 1976),
72. A lighter biography than Allen's, but full of Morgan family and laboratory stories.

**in the shape of a trident:** For a detailed account of this episode, see
Kohler, *Lords*, 39–43.

**"There's two years' work wasted":** R. G. Harrison, "Embryology and Its
Relations," *Science* 85 (1937): 369–74.

23 **"how is the white-eyed fly?":** A Morgan family legend. The story gets the
chronology wrong (the baby was born months before the white-eyed fly),
but it does capture the feeling of the Morgans' *annus mirabilis*. The detail
of the jar next to the bed comes from Shine and Wrobel, *Morgan*, 66.

**Morgan paired the white-eyed fly:** T. H. Morgan, "Sex Limited Inheritance in *Drosophila*," *Science* 32 (1910): 120–2; T. H. Morgan, "The Origin
of Five Mutations in Eye Color in *Drosophila* and Their Modes of Inheritance," *Science* (1911) 33: 534–7; T. H. Morgan, "Genesis of the White-
eyed Mutant," *Journal of Heredity* 33 (1942): 91–2.

24 **"wild over chromosomes":** Quoted in Allen, *Morgan*, 131.

**"Tell it straight":** S. Benzer, personal communication.

26 **"follows the same scheme":** Quoted in A. H. Sturtevant, "Thomas Hunt
Morgan," *Biographical Memoirs of the National Academy of Sciences USA*
33 (1959): 283–325.

**Now Morgan experimented:** See A. H. Sturtevant, "The Linear Arrangement of Six Sex-linked Factors in *Drosophila*, as Shown by Their Mode of
Association," *Journal of Experimental Zoology* 14 (1913): 43–59; J. F. Crow,
"A Diamond Anniversary: The First Chromosome Map," *Genetics* 118
(1988): 1–3.

28 **the idea of his life:** Sturtevant tells the story in his *History of Genetics*
and in his paper "Linear Arrangement."

30 **inherited both** *white* **and** *vermilion:* They could tell this by further breeding experiments. In the current jargon, they had to "progeny test" each fly.

31 **"I had quite a lot":** Garland E. Allen, unpublished interview with A. H. Sturtevant, July 24, 1965, p. 28, Caltech Archives.

**"one of the most":** Quoted in Shine and Wrobel, *Morgan,* 92.

CHAPTER THREE: WHAT IS LIFE?

33 **"With all his amateurish":** Sinclair Lewis, *Arrowsmith* (New York: Harcourt Brace Jovanovich, 1990 [1925]), 316.

**"He was a revolutionary":** Quoted in Robert P. Crease and Charles C. Mann, *The Second Creation: Makers of the Revolution in Twentieth-Century Physics* (New York: Macmillan Publishing Company, 1986), 25.

34 **"I remember the 'awe-full' moment":** Stern, "Continuity of Genetics," 899.

**"You had to keep working on it":** Allen, interview with Sturtevant, 18.

**to help unite biology:** T. H. Morgan, "The Relation of Biology to Physics," *Science* 65 (1927): 213–20.

**something of the general:** For more of the Morgan family tree, see Sturtevant, "Thomas Hunt Morgan."

35 **Nobel address:** T. H. Morgan, "The Relation of Genetics to Physiology and Medicine," *Scientific Monthly* 41 (1935): 5–18. Morgan presented his Nobel lecture in Stockholm on June 4, 1934.

**"What are genes?":** Ibid., 7–8.

**Bensonhurst:** I drew here on interviews with Benzer and members of his family, and on Aspaturian's interviews.

37 **"He had worn the threadbare top-coat":** Lewis, *Arrowsmith,* 11.

**"tyrannical honesty":** Ibid., 119.

**"I make many":** Ibid., 301.

**"never ventured on original experiments":** Ibid., 17.

**"that dead-black spider-web script":** Ibid., 297.

**short, high-strung, nervous:** For a sketch of Muller's personality, see T. Mohr, "Hermann J. Muller, 1890–1967," *Journal of Heredity* 63 (1972): 132–4.

**"The rest of us":** Allen, interview with Sturtevant, 26.

38 **bombarding the flies:** H. J. Muller, "Artificial Transmutation of the Gene," *Science* 66 (1927): 84–7; H. J. Muller, "The Production of Mutations by X-rays," *Proceedings of the U.S. National Academy of Sciences* 14 (1928): 714–26.

**Max Delbrück:** Besides interviews with some of Delbrück's friends, colleagues, and family members, my chief sources included the biography by Ernst Peter Fischer and Carol Lipson, *Thinking About Science: Max Delbrück and the Origins of Molecular Biology* (New York: W. W. Norton & Company, 1988). I also drew on the memoirs in John Cairns, Gunther S. Stent, et al., eds., *Phage and the Origins of Molecular Biology* (Cold Spring Harbor, N.Y.: Cold Spring Harbor Laboratory of Quantitative Biology, 1966), and on a colorful series of interviews with Delbrück conducted by Carolyn Harding in 1979 as part of Caltech's in-house oral history project.

**"Forbidding-looking papers":** Ibid., 63.

39 **the label of a fly bottle:** I took this example from Ralph J. Greenspan, *Fly Pushing: The Theory and Practice of Drosophila Genetics* (Cold Spring Harbor, N.Y.: Cold Spring Harbor Laboratory Press, 1997), 7–11.

40 **"a small boy announces his presence":** M. Delbrück and M. B. Delbrück, "Bacterial Viruses and Sex," *Scientific American* 179 (1948): 49.

**elegantly simple experiments:** Probably the best and certainly the liveliest history of the birth of molecular biology is Horace Freeland Judson, *The Eighth Day of Creation* (New York: Simon and Schuster, 1979). For others, see Robert C. Olby, *The Path to the Double Helix* (Seattle: University of Washington Press, 1974); G. S. Stent and R. Calendar, *Molecular Genetics: An Introductory Narrative* (San Francisco: W. H. Freeman and Company, 1978); and Cairns et al., *Phage.*

**"They must come together":** Morgan, "Relation of Biology to Physics," 216.

**A key discovery:** O. T. Avery, C. M. Macleod, et al., "Induction of Transformation by a Desoxyribonucleic Acid Fraction Isolated from Pneumococcus Type III," *Journal of Experimental Medicine* 79 (1944): 137–58.

**"He just didn't feel":** Allen, interview with Sturtevant, 26.

**a secret wartime project:** My account is based on my own interviews with Benzer, and Aspaturian's. Of several book-length histories of the early days of electronics, I have found only one that tells the story of Benzer's contribution; see Michael Eckert and Helmut Schubert, *Crystals, Electrons, Transistors: From Scholar's Study to Industrial Research,* Thomas Hughes, trans. (New York: American Institute of Physics, 1989), 146–8.

42 **"The electron!":** See, e.g., Crease and Mann, *Second Creation,* 22.

**passed him a book:** Erwin Schrödinger, *What Is Life? And Mind and Matter* (Cambridge, England: Cambridge University Press, 1967 [1944]). Schrödinger was apparently unaware of Herman Muller's prophetic lectures and essays on the same subject. See H. Muller, "Physics in the Attack on the Fundamental Problems of Genetics," *Scientific Monthly* 44 (1936): 210–14. See also E. A. Carlson, "An Unacknowledged Founding of Molecular Biology: H. J. Muller's Contribution to Gene Theory, 1910–1936," *Journal of the History of Biology* 4 (1971): 149–70.

43 **"We seem to arrive":** Schrödinger, *What Is Life?,* 87.

**"But please":** Ibid., 91

44 **"almost absolute":** Ibid., 32–3.

**Admiralty Headquarters:** Francis Crick, *What Mad Pursuit: A Personal View of Scientific Discovery* (New York: Basic Books, 1988), 15–18. Gunther Stent, another young physicist who converted, describes the romance of reading *What Is Life?* in André Lwoff and Agnes Ullmann, eds., *Origins of Molecular Biology: A Tribute to Jacques Monod* (New York: Academic Press, 1979), 232. The influence of the book and that of physicists themselves became part of the legend of molecular biology. The mantle and authority of physics helped establish the new science. See Evelyn Fox Keller, "Physics and the Emergence of Molecular Biology: A History of Cognitive and Political Synergy," *Journal of the History of Biology* 23 (1990): 389–409.

**"from the moment":** James D. Watson, "Growing Up in the Phage Group," in Cairns et al., *Phage,* 239.

"*I* **work on viruses**": I reconstructed this scene using my interviews with Benzer, and Aspaturian's.

CHAPTER FOUR: THE FINGER OF THE ANGEL

46  **"I study myself"**: Michel de Montaigne, "Of Experience." *The Essays of Michel de Montaigne,* George B. Ives, trans. (New York: Heritage Press, 1946 [1580–1595]), vol. 2, 1465.
"**pretty and witty**": Judson, *Eighth Day,* 340.

47  **"He used to really"**: Aspaturian interviews, 79.
**a small number of signs**: Schrödinger, *What Is Life?,* 65–6.
**writing a report**: J. D. Watson and F. H. C. Crick, "A Structure for Deoxyribose Nucleic Acid," *Nature* 171 (1953): 737–8.
"**suffered from periodic fears**": Francis Crick, "The Double Helix: A Personal View," *Nature* (1974): 766–71.

49  **call itself molecular biology**: But the name had been suggested years before; see Warren Weaver, "Molecular Biology: Origin of the Term," *Science* 170 (1970): 581–2.
**Benzer hit on a plan**: He was building on a suggestion presented by G. Pontecorvo in "Genetic Formulation of Gene Structure and Gene Action," *Advances in Enzymology* 13 (1952): 121–49.

50  **A few of Morgan's Raiders**: This was Pontecorvo's argument, for example.

51  **Benzer's plan**: For a few of his own accounts of the experiment, see S. Benzer, "The Structure of a Genetic Region in Bacteriophage," *Proceedings of the U.S. National Academy of Sciences* 41 (1955): 344–54; S. Benzer, "Genetic Fine Structure," Harvey Lectures 56 (1960): 1–21; S. Benzer, "The Fine Structure of the Gene," *Scientific American* (January 1962): 2–15; and S. Benzer, "Adventures in the rII Region," in Cairns et al., *Phage,* 157–65. Gunther Stent devotes a chapter of his history of molecular biology to Benzer's experiment: "Genetic Fine Structure," in *Molecular Genetics,* 375–412. My account draws on Aspaturian's interviews, and on my own interviews with Benzer and other molecular biologists who watched the *rII* experiment unfold.

53  **"Delusions of grandeur"**: Benzer, "Adventures," 162.
"**You must have drunk**": Aspaturian interviews, 101.
"**One can . . . perform**": Benzer, "The Fine Structure of the Gene," 2.
"**There was no way**": Judson, *Eighth Day,* 274.
"**Then his research wiped out**": Lewis, *Arrowsmith,* 330.

54  **the Kapitza Club**: Aspaturian interviews, 135.
**The discovery of the neutron**: Richard Rhodes, *The Making of the Atomic Bomb* (New York: Simon & Schuster, 1986), 165.
**a single yes or no**: Walter Gratzer, ed., *A Literary Companion to Science* (New York: W. W. Norton & Company, 1989), 171.
**Hershey Heaven**: Galton envied Mendel and Darwin for finding a kind of Hershey Heaven. In his memoir he notes that they had each turned the world upside down while "never or hardly ever" leaving their homes. Galton, *Memories,* 308. Today many molecular biologists dream of Her-

shey Heaven; see Robert Pollack, "A Crisis in Scientific Morale," *Nature* 385 (1997): 673–4.

55 **a retrospective volume:** James A. Peters, ed., *Classic Papers in Genetics,* Prentice-Hall Biological Science Series (Englewood Cliffs, N.J.: Prentice-Hall, 1959).

56 **happened to mention the word:** Aspaturian interviews, 152.

57 **began collecting typographical errors:** Benzer, "Genetic Fine Structure," 2–3.

58 **loved Benzer's tricks:** Richard P. Feynman, *Surely You're Joking, Mr. Feynman!* (New York: Bantam Books, 1986 [1985]), 59–63. James Gleick, *Genius: The Life and Science of Richard Feynman* (New York: Pantheon Books, 1992), 349–51.

"**finding one man in China":** Ibid., 350.

**Crick and Sydney Brenner:** Crick devotes a chapter of his memoir to this work: "Triplets," in Crick, *Mad Pursuit,* 122–36.

**a meeting in India:** Aspaturian interviews, 115.

**In Paris:** For eloquent accounts of this work, see François Jacob, *The Statue Within,* F. Philip, trans. (New York: Basic Books, 1988); and F. Jacob, *The Logic of Life: A History of Heredity,* Betty E. Spillmann, trans. (Princeton, N.J.: Princeton University Press, 1993).

**Jewish legend:** Talmud Bavli, Niddah 30b.

59 **an atomic force microscope:** Erik Stokstad, "DNA on the Big Screen," *Science* 275 (1997): 1882.

60 **a flock of sheep:** I. Wilmut et al., "Viable Offspring Derived from Fetal and Adult Mammalian Cells," *Nature* 385 (1997): 810–13.

"**Every day, at lunch":** Jacob, *Statue,* 261.

CHAPTER FIVE: A NEW STUDY, AND A DARK CORNER

61 "**Psychology was to him":** Henry Adams, *The Education of Henry Adams* (Boston: Houghton Mifflin, 1973 [1907]), 231. Quoted in Gerald M. Edelman, *Bright Air, Brilliant Fire: On the Matter of the Mind* (London: Penguin Books, 1992), 33.

"**It is still":** Rhodes, *Atomic Bomb,* 11.

"**the borderline between":** Crick, *Mad Pursuit,* 17.

62 "**the hubris of the physicist":** Ibid., 13.

"**Don't worry, Ducky":** Ibid., 9.

**there was nothing left:** See, e.g., Gunther S. Stent, "That Was the Molecular Biology That Was," *Science* 160 (1968): 390–5.

"**that's my cousin Seymour!":** Aspaturian interviews, 154.

"**He wanted my wife":** Ibid., 132.

**After the memoir was published:** James D. Watson, *The Double Helix: A Personal Account of the Discovery of the Structure of DNA* (New York: Atheneum, 1968). Gunther S. Stent has edited a critical edition containing commentaries, reviews, and reprints of some of the original scientific papers (New York: W. W. Norton & Company, 1980).

63 "**I confess":** Crick, *Mad Pursuit,* 145.

*The Double Helix* **replaced** *Arrowsmith:* See, e.g., the Watson portrait in E. O. Wilson's memoir *Naturalist* (Washington, D.C.: Island Press,

1994), 224: "Watson was a boy's hero of the natural sciences, the fast young gun who rode into town."

**Crick told one historian:** Judson, *Eighth Day,* 193.

**"SCIENTISTS ARE HUMAN":** Crick, *Mad Pursuit,* 83.

**behavior of single cells: Fischer and Lipson tell the story in** *Thinking About Biology,* 234–45. See also E. Cerdá-Olmedo and E. D. Lipson, eds., *Phycomyces* (Cold Spring Harbor, New York: Cold Spring Harbor Laboratory, 1987); R. K. Clayton and M. Delbrück, "Purple Bacteria," *Scientific American* 185, no. 11 (1951): 68–72; and M. Delbrück, "Primary Transduction Mechanisms in Sensory Physiology and the Search for Suitable Experimental Systems," *Israel Journal of Medical Science* 1 (1965): 1363–65.

64  **dictated a letter:** Fischer and Lipson, *Thinking About Biology,* 234.

65  **"Are we doing things":** Aspaturian interviews, 165.

**"When you have one child":** S. Benzer, Crafoord Prize Lecture, Stockholm, September 27, 1993.

**"gossip test":** Crick, *Mad Pursuit,* 17.

**"He was right":** Benzer, "Adventures," 157.

**"I had almost gone":** Ibid., 165.

66  **"a healthy corrective":** Crick, *Mad Pursuit,* 14.

**"I think, but he knows":** Fischer et al., *Thinking,* 133.

**"I must beg you":** Freud, "Lay Analysis," in Stevenson, *Human Nature,* 167.

**"We must recollect":** Freud, "On Narcissism"; quoted in Melvin Konner, *The Tangled Wing: Biological Constraints on the Human Spirit* (New York: Holt, Rinehart and Winston, 1982), xv.

**full of cotton:** See M. E. Bitterman, "Psychology via Physiology: Review of *The Neuroscience of Animal Intelligence,* by Euan M. Macphail," *Science* 263 (1994): 1635–6.

67  **"the stifling soul cloud":** J. B. Watson, Chap. 1, *Behaviorism* (New York: W. W. Norton, 1925); excerpted in Stevenson, *Human Nature,* 193–8.

**"guarantee to take any one":** Watson, *Behaviorism,* 81; quoted in Carl Degler, *In Search of Human Nature* (Oxford: Oxford University Press, 1991), 81.

**"Very few people":** B. F. Skinner, *Science and Human Behavior* (New York: Macmillan, 1953); excerpted in Stevenson, *Human Nature,* 199–218.

**without a psyche:** Paul F. Cranefield uses this phrase in a somewhat different context in "The Philosophical and Cultural Interests of the Biophysics Movement in 1847," *Journal of the History of Medicine and Allied Sciences* 21 (1966): 7.

**worn down to stubs:** Gary Cziko, *Without Miracles: Universal Selection Theory and the Second Darwinian Revolution* (Cambridge, Mass.: MIT Press, 1995), 117.

**"The present unhappy condition":** Skinner, *Science and Human Behavior;* excerpted in Stevenson, *Human Nature,* 204.

68  **"knowledge-microscopists":** Nietzsche, *Beyond Good and Evil;* excerpted in Monroe C. Beardsley, ed., *The European Philosophers from Descartes to Nietzsche* (New York: The Modern Library, 1993), 810.

**"For precept must be":** Isaiah 28: 10, *The Reader's Bible* (New York: Oxford University Press, 1951), 943.

69 **"visibly shift"**: Plato, *The Last Days of Socrates* (Harmondsworth, England: Penguin Books, 1959), 40.

**"We burn with desire"**: Blaise Pascal, *Pensées* (London: Penguin Books, 1995), 63.

**"To study Metaphysics"**: Darwin, *Metaphysics,* 71.

70 **"Round and round"**: D. H. Lawrence, "Man and Bat," in *Birds, Beasts, and Flowers* (New York: Haskell House, 1974), 103. Quoted in T. Tanner, "Out of England: Review of *D. H. Lawrence: Triumph to Exile, 1912–1922,* by Mark Kinkead-Weekes," *Times Literary Supplement* 4873 (1996): 3–4.

### PART TWO: KONOPKA'S LAW

71 **"Things are always best"**: Pascal, *Lettres Provinciales;* quoted in John Bartlett, *Familiar Quotations,* 14th ed. (Boston: Little, Brown and Company, 1968), 363.

### CHAPTER SIX: FIRST LIGHT

73 **"I'll tell you how the Sun rose"**: Emily Dickinson, *The Complete Poems of Emily Dickinson* (Boston: Little, Brown and Company, 1960), 150.

**"Everyone who ever lived"**: Richard Powers, *Galatea* 2.2 (New York: HarperCollins, 1995), 8.

**a little book**: Dean E. Wooldridge, *The Machinery of the Brain* (New York: McGraw-Hill Book Company, 1963).

**cut the optic nerves**: Ibid., 20–2.

74 **experiments with cats**: Ibid., 169–74.

**repeated this experiment**: Ibid., 181.

75 **"a travelogue"**: Ibid., vii.

76 **Cutting the corpus callosum**: Eric R. Kandel, James H. Schwartz, and Thomas M. Jessell, eds., *Principles of Neural Science* (Norwalk, Conn.: Appleton & Lange, 1991), 833.

**dominant for action and movement**: Ibid., 835.

**"I'm going to get a Coke"**: Richard M. Restak, *The Mind* (New York: Bantam Books, 1988), 27.

**its own characteristic function**: Kandel, *Principles,* 7–9.

77 **"I am a parcel"**: Henry David Thoreau, "Sic Vita," in F. O. Matthiessen, ed., *The Oxford Book of American Verse* (New York: Oxford University Press, 1950), 241.

**parliament of instincts**: Konrad Lorenz, *On Aggression,* Marjorie Kerr Wilson, trans. (New York: Bantam Books, 1966), 81.

**"Humans were ruled out"**: Benzer, Crafoord Lecture.

78 **"Is astonishment expressed"**: Darwin, *Expression,* 15.

**"I put my face"**: Ibid., 38.

79 **"I have never tried"**: Charles Darwin, *The Life and Letters of Charles Darwin,* Francis Darwin, ed. vol. 3 (London, 1887), 238; quoted in Janet Browne, *Charles Darwin,* vol. 1 (New York: Alfred A. Knopf, 1995), xii.

"I think that a delight": Francis Galton, *Inquiries into Human Faculty and Its Development* (London: Macmillan and Co., 1883), 87.

a calculating prodigy: Galton, *Memories,* 271.

"up went a multitude": Ibid., 273.

"the number of operations": Galton, *Inquiries,* 186.

into the basement: Ibid.

80 "I myself have": Ibid., 58–9.

81 "but I have seen": Ibid., 60.

82 "Are they out of their minds?": Wilson, *Naturalist,* 219–20.

"Imagine: biology transformed": Ibid., 44.

"Without a trace of irony": Ibid., 218–19.

"solemn, arrogant": Peter Gay, *The Enlightenment: An Interpretation* (New York: Alfred A. Knopf, 1966), 16.

83 "He was a nice guy": Aspaturian interviews, 117.

84 mixed up: Konrad Z. Lorenz, *King Solomon's Ring* (New York: Signet, 1972 [1952]), xvi.

"Don't do fashionable research": Fischer and Lipson, *Thinking About Biology,* 235.

CHAPTER SEVEN: First Choice

86 "The brain is so vigorous": Nathaniel Wanley, *The Wonders of the Little World* (London, 1788), 6; quoted in Leonard Barkan, *Nature's Work of Art: The Human Body as Image of the World* (New Haven: Yale University Press, 1975), 34.

"I thought of a maze": Jorge Luis Borges, "The Garden of Forking Paths," in *Labyrinths: Selected Stories and Other Writings,* Donald A. Yates and James E. Irby, trans. (New York: New Directions, 1964), 23.

87 "I laid out": Aspaturian interviews, 183.

"Obviously, that attitude": Ibid., 185.

89 "At faculty meetings": Wilson, *Naturalist,* 221.

"strange, though mistaken": Charles Darwin, *The Descent of Man, and Selection in Relation to Sex* (Princeton, N.J.: Princeton University Press, 1981 [1871]), 43.

"What sort of insects": Lewis Carroll, *Alice's Adventures in Wonderland and Through the Looking-Glass* (New York: New American Library, 1960 [1865 and 1871]), 151.

"Remember, there is": Lwoff, *Origins,* 136.

"Dr. Horovitz, this is": Aspaturian interviews, 38.

90 houses he had seen in Italy: Galton, *Natural Inheritance,* 8.

91 "Suppose we were building": Ibid.

92 "namely, good in stock": Galton, *Inquiries,* 24.

"We must free our minds": Ibid., 3.

"capricious and coy": Ibid., 56.

double-dealing misers: Galton once warmed up a lecture audience of fellow scientists at the British Association with a joke about stingy Jews; see Galton, *Memories,* 272.

"A little goodwill": T. H. Morgan, *Evolution and Genetics,* 2d ed. (Princeton, N.J.: Princeton University Press, 1925), 206–7; quoted in Daniel J.

Kevles, *In the Name of Eugenics,* 2d ed. (Cambridge, Mass.: Harvard University Press, 1995), 133.

92  **ardent eugenicist:** See, e.g., H. J. Muller, *Out of the Night: A Biologist's View of the Future* (New York: Garland Publishing, 1984 [1935]).
**transformation of the human species:** The last sentence is full of suppressed excitement: "The time is not ripe to discuss here such possibilities with reference to the human species." Muller, "Artificial Transmutation," 87.
**"a whole genus":** Galton, *Memories,* 175.

93  **"But it does not":** Galton, *Inquiries,* 27.
**"Health Fair":** Lewis, *Arrowsmith,* 268–72.
**helped inspire the Nazis:** Diane B. Paul, *Controlling Human Heredity: 1865 to the Present* (Atlantic Highlands, N.J.: Humanities Press, 1995), 86.
**keynote speech:** The Galton Lecture, delivered before the Eugenics Society in London on February 17, 1936; in Julian S. Huxley, "Eugenics and Society," *The Eugenics Review* 28, no. 1 (1936): 11–35.
**"one of the supreme":** Ibid., 11.
**"We cannot digest":** Ibid., 33–4.

94  **"In some sort":** J. R. Oppenheimer, "Physics in the Contemporary World," lecture at Massachusetts Institute of Technology, November 25, 1947; quoted in Bartlett, *Familiar Quotations,* 1055.
**"Such a change":** Rhodes, *Atomic Bomb,* 26.
**"A girl of sixteen":** C. P. Blacker, " 'Eugenic' Experiments Conducted by the Nazis on Human Subjects," *Eugenics Review* 44, no. 1 (1952): 11.
**"A man must be":** Galton, *Natural Inheritance,* 155.

95  **"nothing but the form":** Quoted in Fred H. Wilhoite, Jr., "Ethology and the Tradition of Political Thought," *Journal of Politics* 33 (1971): 628–9; quoted in Degler, *Human Nature,* 230.
**"Other people want":** Harding, interviews with Delbrück, 88.

96  **"Oh, I don't know":** Aspaturian interviews, 48–9.
**"no decorative heroisms":** Lewis, *Arrowsmith,* 333.

CHAPTER EIGHT: FIRST TIME

98  **"—As if the idea":** Darwin, *Metaphysics,* 8.

99  **"The work can be done":** T. H. Morgan, letter to E.B. Babcock, June 15, 1920; quoted in Allen, *Morgan,* 291.

100  **flight tester:** Benzer, "Genetic Dissection of Behavior," 28.

101  **"And you claim":** Freud, "Lay Analysis," in Stevenson, *Human Nature,* 166.
*mad* **in honor of Max:** K. Bergmann, A. P. Eslava, and E. Cerdá-Olmedo, "Mutants of *Phycomyces* with Abnormal Phototropism," *Molecular and General Genetics* 123 (1973): 1–16; Fischer and Lipson, *Thinking About Biology,* 249. For a review of this and other early work in the atomic theory of behavior, see William G. Quinn and James L. Gould, "Nerves and Genes," *Nature* 278 (1979): 19–23.
**"trying all kinds":** Fischer and Lipson, *Thinking About Biology,* 245.

102  **his first paper:** Benzer, "Behavioral Mutants," 1967.
**"And doubtless":** Quoted in Fraser, *Of Time,* 179.

"The sensitive plant": De Mairan, "Observation Botanique," *Histoire de l'Académie Royale des Sciences* (Paris, 1729), 35; quoted in Ritchie R. Ward, *The Living Clocks* (New York: Alfred A. Knopf, 1971), 44–5. Ward is the chief source of my description of early clock work.

103 The passion flower would open: Fraser, *Of Time,* 182.
*Euglena* swims like an animal: Arthur T. Winfree, *The Timing of Biological Clocks* (New York: Scientific American Books, 1987), 111.

104 "All this showing": Spinoza, *Ethics;* excerpted in Stevenson, *Human Nature,* 94–5.
elaborate experiments involving carrots: Ward describes Brown's work in *Living Clocks,* 259–78.
South Pole: Ibid., 279–99.

105 checked the clock in fruit flies: Colin S. Pittendrigh, "Temporal Organization: Reflections of a Darwinian Clock-watcher," *Annual Review of Physiology* 55 (1993): 17–54.
"What we need": Brown, *Living Clocks,* 299.
round-the-clock watches: Pittendrigh, "Temporal Organization."

111 wrote up a report: Ronald J. Konopka and Seymour Benzer, "Clock Mutants of *Drosophila melanogaster," Proceedings of the U.S. National Academy of Sciences* 68 (1971): 2112–16.
"I don't believe it": I reconstructed this scene from interviews with Benzer and Konopka and from the account in Greenspan, "Emergence of Neurogenetics," 150.

CHAPTER NINE: FIRST LOVE

112 "What is it men": "The Question Answer'd," *Portable Blake,* 135.
"embrangled in inextricable difficulties": Quoted in Fraser, *Of Time,* 11.
try to put it into words: Saint Augustine, *Confessions,* Book 11: quoted in James McConkey, ed., *The Anatomy of Memory* (New York: Oxford University Press, 1996), 50.

113 "all the dense fullness": Plotinus, *The Third Ennead;* quoted in Fraser, *Of Time,* 23–4.
"Plato says that": Darwin, *Metaphysics,* 30.
"was known principally": David Park, *The Image of Eternity: Roots of Time in the Physical World* (Amherst, Mass.: The University of Massachusetts Press, 1980), 16.
"Three things are": Proverbs 30: 18–19.

114 "No, this trick won't": Richard Powers, *The Gold Bug Variations* (New York: HarperPerennial, 1992), 124.
"When you swim": Roger Payne, *Among Whales* (New York: Scribner's, 1995), 145.
Male bowerbirds: Frank B. Gill, *Ornithology* (New York: W. H. Freeman and Company, 1989), 179–81.

115 "One bower was": Ibid., 180.
nematode worm: For early papers laying out the worm's possibilities, see S. Brenner, "The Genetics of *Caenorhabditis elegans," Genetics* 77 (1974): 71–94; and J. E. Sulston and S. Brenner, "The DNA of *Caenorhabditis*

*elegans," Genetics* 77 (1974): 95–104. Paul Sternberg is one of many molecular biologists now working on behavioral mutants in the worm. See, e.g., Katharine S. Liu and Paul W. Sternberg, "Sensory Regulation of Male Mating Behavior in *Caenorhabditis elegans," Neuron* 14 (1995): 79–89.

116 **In the Hawaiian archipelago:** Herman T. Spieth, "Courtship Behavior in *Drosophila," Annual Review of Entomology* 19 (1974): 385–405; Herman T. Spieth, *Courtship Behaviors in the Hawaiian Picture-winged Drosophila* (Berkeley: University of California Press, 1984).

"**contact of the labellar lobes**": Ibid., 10.

"**If the female responds**": Ibid., 11.

117 "**with the labellar lobes**": Ibid., 12.

119 **Hall borrowed a set of mutants:** They came from the Fly Lab of Dan Lindsley, at UCSD.

120 "**Glory be to God**": Gerard Manley Hopkins, "Pied Beauty," in Louis Untermeyer, ed., *Modern American Poetry. Modern British Poetry* (New York: Burlingame, 1958), 39.

121 **a corkscrewing path:** For a photograph, see Benzer, "Behavioral Mutants," 25.

"**Sometimes**": Ibid., 31.

**Sturtevant had realized:** A. H. Sturtevant, "The Use of Mosaics in the Study of the Developmental Effects of Genes," *Proceedings of the Sixth International Congress of Genetics* (1932): 304–7.

122 **They drew an oval map:** A. Garcia-Bellido and J. R. Merriam, "Cell Lineage of the Imaginal Discs in *Drosophila* Gynandromorphs," *Journal of Experimental Zoology* 170 (1969): 61–75.

**Benzer and Hotta:** Y. Hotta and S. Benzer, "Mapping of Behaviour in *Drosophila* Mosaics," *Nature* 240 (1972): 527–35; Y. Hotta and S. Benzer, "Courtship in *Drosophila* Mosaics: Sex-specific Foci for Sequential Action Patterns," *Proceedings of the U.S. National Academy of Sciences* 73, no. 11 (1976): 4154–8.

"**that was a sentimental thing**": Aspaturian interviews, 232.

**fate maps to explore sexual instincts:** Douglas R. Kankel and Jeffrey C. Hall, "Fate Mapping of Nervous System and Other Internal Tissues in Genetic Mosaics of *Drosophila melanogaster," Developmental Biology* 48 (1976): 1–24; J. C. Hall, "Control of Male Reproductive Behavior by the Central Nervous System of *Drosophila*: Dissection of a Courtship Pathway by Genetic Mosaics," *Genetics* 92 (1979): 437–57.

124 *bithorax . . . Antennapedia*: E. B. Lewis, "A Gene Complex Controlling Segmentation in *Drosophila," Nature* 276 (1978): 565–70; E. B. Lewis, "Clusters of Master Control Genes Regulate the Development of Higher Organisms," *Journal of the American Medical Association* 267 (1992): 1524–31.

125 "**It is as if**": Christopher Wills, *The Wisdom of the Genes* (New York: Basic Books, 1989), 235.

126 **The most surprising courtship mutant:** K. S. Gill, "A Mutation Causing Abnormal Courtship and Mating Behavior in Male *Drosophila melanogaster," American Zoologist* 3 (1963): 507.

CHAPTER TEN: FIRST MEMORY

128 **"Memory is a passion"**: Elie Wiesel, *All Rivers Run to the Sea* (New York: Alfred A. Knopf, 1995), 150.

129 **"inspiring spiritual atmosphere"**: Stern, "Continuity of Genetics," 906.

130 **"Tell me"**: W. Blake, "Visions of the Daughters of Albion," *Portable Blake*, 292. Blake goes on to ask, "Tell me where dwell the thoughts forgotten till thou call them forth?"

131 **a famous memoir**: A. R. Luria, *The Mind of a Mnemonist*, Lynn Solotaroff, trans. (New York: Avon Books, 1969).
**backward, forward, or even diagonally**: Ibid., 17.
**"I simply had to admit"**: Ibid., 11.

132 **read the *Encyclopaedia Britannica***: E. B. Lewis, "Remembering Sturtevant." *Genetics* 41 (1995): 1227–30.
**"For us to learn"**: Yadin Dudai, *The Neurobiology of Memory: Concepts, Findings, Trends* (Oxford: Oxford University Press, 1989), 3.
**"a rope let down from heaven"**: Marcel Proust, *Remembrance of Things Past* (New York: Random House, 1981) Vol. III: 912; quoted in Stephen S. Hall, "Our Memories, Ourselves," *The New York Times Magazine* (February 15, 1998): 26.

133 **"a looking-glass"**: Locke, *Human Understanding*, Book 2, Chapter 1, Section 15, p. 65.

134 **"little machines in a deep sleep"**: V. G. Dethier, "Microscopic Brains," *Science* 143 (1964): 1138–45. See also Dethier's charming book *To Know a Fly* (San Francisco: Holden-Day, 1962).
**"Can't learn anything"**: Howard Simons, "Scientist Finds Flies Can't Learn But Moths and Bats Use Sonar," *Washington Post*, April 28, 1966. " 'You name it and we've tried it,' " Dethier said dejectedly yesterday at the National Academy of Sciences. 'But nothing works.' Flies are incapable of learning. Now Dethier is trying caterpillars."

135 **"And at some level"**: My description of Quinn's early work comes from interviews with Quinn, Benzer, and colleagues, and from their first paper: William G. Quinn, William A. Harris, and Seymour Benzer, "Conditioned Behavior in *Drosophila melanogaster*," *Proceedings of the U.S. National Academy of Sciences* 71, no. 3 (1974): 708–12.

138 **"A single experience"**: Schrödinger, *Mind and Matter*, 102.
**"Chip Quinn once"**: S. Benzer, "A Fly's Eye View of Development," lecture, Symposium on Molecular Biology of Development, Cold Spring Harbor Laboratory, Cold Spring Harbor, N.Y., May 29, 1985.
**"A worm is only a worm"**: Denis Diderot, *Rameau's Nephew and Other Works*, Jacques Barzun and Ralph H. Bowen, trans. (Indianapolis, Ind.: The Bobbs-Merrill Company, 1964), 119.

139 **no talent for learning**: Yadin Dudai, Yuh-Nung Jan, et al., "*dunce*, a Mutant of *Drosophila* Deficient in Learning," *Proceedings of the U.S. National Academy of Sciences* 73 (1976): 1684–8.

140 **"The body is but a watch"**: Julien Offray de La Mettrie, *Machine Man and Other Writings*, Ann Thomson, trans. and ed. (Cambridge: Cambridge University Press, 1996), 31.
**"The selfish gene"**: Richard Dawkins, *The Selfish Gene*, 2d ed. (Oxford: Oxford University Press, 1989).

140 "A man is but what he knoweth": Quoted in René Dubos, *So Human an Animal* (New York: Charles Scribner's Sons, 1968), 111.

### PART THREE: PICKETT'S CHARGE

143 "and thus beneath": Conrad Aiken, "Time in the Rock, or Preludes to Definition," *Selected Poems* (Cleveland, Ohio: Meridian Books, 1964), 148.

### CHAPTER ELEVEN: THE DROSOPHILA ARMS

145 "The flies, poor things": "The Invisible World," in Primo Levi, *Other People's Trades* (New York: Summit Books, 1989), 60.
In the scientific literature: Many of the classic studies of *Drosophila* courtship had been done by Aubrey Manning and colleagues at the University of Edinburgh. For a review, see Aubrey Manning, "*Drosophila* and the Evolution of Behaviour," *Viewpoints in Biology* 4 (1964): 125–69.
So Kyriacou built a recording studio: My account of Kyriacou's early work comes from interviews with Kyriacou, Hall, and colleagues, and from C. P. Kyriacou and Jeffrey C. Hall, "Circadian Rhythm Mutations in *Drosophila melanogaster* Affect Short-Term Fluctuations in the Male's Courtship Song," *Proceedings of the U.S. National Academy of Sciences* 77, no. 11 (1980): 6729–33.

147 "Triple phrases sound as to a drumbeat": Garry Wills, *Lincoln at Gettysburg: The Words That Remade America* (New York: Simon & Schuster, 1992), 171–72.

149 "*Mein Gott!*": The phage watcher was J. J. Bronfenbrenner. His surprise is described in Thomas F. Anderson, "Electron Microscopy of Phages," in Cairns et al., *Phage,* 65.

151 a jumping gene called *mariner*: T. Oosumi, W. R. Belknap, and B. Garlick, "*Mariner* Transposons in Humans," *Nature* 378 (1995): 672.

152 "A Valentine for NIH": M. Delbrück, *Trends in Biochemical Science* (February 1980): xii; quoted in Fischer and Lipson, *Thinking About Biology,* 283.

### CHAPTER TWELVE: CLONING AN INSTINCT

155 "No doubt the process": Marcel Proust, *Time Regained,* vol. 3 of *Remembrance of Things Past,* C. K. Scott-Moncrieff and Terence Kilmartin, trans. (New York: Random House, 1982), 912.
the giant chromosomes: Their discovery had been a major step for geneticists. See, e.g., Theophilus S. Painter, "Salivary Chromosomes and the Attack on the Gene," *Journal of Heredity* 25 (1934): 464–76.

159 raced their papers into print: William A. Zehring et al., "P-Element Transformation with *period* Locus DNA Restores Rhythmicity to Mutant, Arrhythmic *Drosophila melanogaster*," *Cell* 39 (1984): 369–76; T. A. Bargiello, F. R. Jackson, and M. W. Young, "Restoration of Circadian Behavioural Rhythms by Gene Transfer in *Drosophila,*" *Nature* 312 (1984): 752–4.

160 **jumped from one species to another:** David A. Wheeler et al., "Molecular Transfer of a Species-Specific Behavior from *Drosophila simulans* to *Drosophila melanogaster*," *Science* 251 (1991): 1082–5.

163 **"Jimmy Crack Corn":** The song was written by Bill Wood and sung by Robert Sinsheimer as part of an evening of homegrown theatricals entitled "I Am Curious, Max" at Caltech on November 22, 1969.

164 **"All of you neurogeneticists":** J. C. Hall, "Pleiotropy of Behavioral Genes," in R. J. Greenspan and C. P. Kyriacou, eds. *Flexibility and Constraint in Behavioral Systems* (New York: John Wiley & Sons, 1994), 15–27.
**elegiacal essays:** See, e.g., Gunther S. Stent, "That Was the Molecular Biology That Was," *Science* 160 (1968): 390–5.
**Stent had now reconsidered:** Gunther S. Stent, "Strength and Weakness of the Genetic Approach to the Development of the Nervous System," *Annual Review of Neurosciences* 4 (1981): 163–94.
**"The method of nature":** Ralph Waldo Emerson, "The Method of Nature," in *Essays and Lectures,* 119.
**"sick at heart":** M. Delbrück, diary entry, July 29, 1972; quoted by Fischer and Lipson in *Thinking About Biology,* 251. They write, "This is one of the few emotional comments to appear in the diary, which he kept to the end of his life."

165 **commencement address:** M. Delbrück, "The Arrow of Time—Beginning and End," Caltech, June 9, 1978.

CHAPTER THIRTEEN: READING AN INSTINCT

167 **"I am a book":** Delmore Schwartz, "I Am a Book I Neither Wrote Nor Read," in Nancy Sullivan, ed., *The Treasury of American Poetry* (Garden City, N.Y.: Doubleday & Company, 1978), 548.
**a little handbook:** F. M. Cornford, *Microcosmographia Academica: Being a Guide for the Young Academic Politician* (Cambridge, England: Bowes & Bowes, 1908).
**"the merest sketch":** Ibid., 3.
**a lonely letter:** S. Benzer to Max Delbrück, September 26, 1952, Caltech Archives.

168 **One of its formative meetings:** James Watson, "The Human Genome Initiative," in *Genetics and Society,* Barry Holland and Charabambos Kyriacou, eds. (Reading, Mass.: Addison-Wesley Publishing Company, 1993), 15.

169 **Entrepreneurs raced:** For a sketch of the big science and big business of gene mapping at the close of the twentieth century, see Jon Cohen, "The Genomics Gamble," *Science* 275 (1997): 767–72.
**"I just sold one hundred thousand genes":** Ibid., 769.
**"In genetics there is no mystery":** The molecular biologist is Daniel Cohen of the French biotechnology company Genset; quoted in Michael Balter, ". . . And a Recent Recruit," *Science* 275 (1997): 773.

172 **wrinkled or smooth:** J. R. S. Fincham, "Mendel—Now Down to the Molecular Level," *Nature* 343 (1990): 208–9.
**tall or short:** Stewart B. Rood et al., "Why Mendel's Peas Came Up Short," *Science* 277 (1997): 1611. Rood et al., "Gibberellins: A Phytohormonal Basis for Heterosis in Maize," *Science* 241 (1988): 1216–18.

173　the complete sequence: Young's lab reported a full sequence in March 1986; see F. R. Jackson, T. A. Bargiello, S.–H. Yun, et al., "Product of *per* Locus of *Drosophila* Shares Homology with Proteoglycans," *Nature* 320 (1986): 185–8. Rosbash then published his first sequence, which was partial and focused on the Thr-Gly repeat region, in July 1986; see P. Reddy, A. C. Jacquier, N. Abovich, et al., "The *period* Clock Locus of *D. melanogaster* Codes for a Proteoglycan," *Cell* 46 (1986): 53–61. One year later, Rosbash came out with a paper reporting the full-length sequence of *period* together with the *per-zero* and *per-short* mutations; see Qiang Yu et al., "Molecular Mapping of Point Mutations in the *period* Gene That Stop or Speed Up Biological Clocks in *Drosophila melanogaster*," *Proceedings of the U.S. National Academy of Sciences* 84 (1987): 784–8. Both rival labs were wrong about the proteoglycan connection.

175　"We used to think": J. Watson, quoted in Leon Jaroff, "The Gene Hunt," *Time* 133, no. 12 (March 20, 1989): 62–7.

CHAPTER FOURTEEN: SINGED WINGS

176　"Philosophy is really Homesickness": *Pollen and Fragments: Selected Poetry and Prose of Novalis*, Arthur Versluis, trans. (Grand Rapids, MI: Phanes Press, 1989), 56. Quoted in Richard Holmes, "Paradise in a Dream," *The New York Review of Books* (July 17, 1997): 4–6.

his famous book: E. O. Wilson, *Sociobiology: The New Synthesis* (Cambridge, Mass.: Harvard University Press, 1975).

"Wilson, you're all wet": Wilson, *Naturalist,* 349.

177　he was depressed: Fischer and Lipson, *Thinking About Biology,* 274.

178　six-page, single-spaced letter: Jerry Hirsch, "Benzer's 'Learning' Claim," September 14, 1979.

Hirsch's manifesto: J. Hirsch, "Behavior Genetics and Individuality Understood," *Science* 142 (1963): 1436–42.

180　"The pursuit is repeated": The struggles of sticklebacks were studied by Niko Tinbergen. Lorenz describes them in *King Solomon's Ring,* 45. Though Benzer dropped the debate with Hirsch, rebuttals came from his students and, later, from their students; see Tim Tully, "Measuring Learning in Individual Flies Is Not Necessary to Study the Effects of Single-Gene Mutations in *Drosophila*: A Reply to Holliday and Hirsch," *Behavior Genetics* 16, no. 4 (1986): 449–55.

　　The clash between Hirsch and Benzer was not only a clash of egos but also of traditions, and Tully came to feel that both are valuable. See Tim Tully, "Discovery of Genes Involved with Learning and Memory: An Experimental Synthesis of Hirschian and Benzerian Perspectives," *Proceedings of the U.S. National Academy of Sciences* 93 (1996): 13460–67.

181　a neurocrystal: Benzer coined the term; see Donald F. Ready, Thomas E. Hanson, and Seymour Benzer, "Development of the *Drosophila* Retina, A Neurocrystalline Lattice," *Developmental Biology* 53 (1976): 217–40. For a well-illustrated overview, see Peter A. Lawrence, *The Making of a Fly: The Genetics of Animal Design* (Oxford: Blackwell Scientific Publications, 1992), 180–94.

*sevenless:* William A. Harris, William S. Stark, and John A. Walker,

"Genetic Dissection of the Photoreceptor System in the Compound Eye of *Drosophila melanogaster*," *Journal of Physiology* 256 (1976): 415–39. The discovery of *sevenless* opened an entire field: it is still used as a model of the way cells communicate in the growing embryo.

182 **monoclonal antibodies:** Shinobu C. Fujita et al., "Monoclonal Antibodies Against the *Drosophila* Nervous System," *Proceedings of the U.S. National Academy of Sciences* 79 (1982): 7929–33.

**a marriage of true minds:** Carol A. Miller and Seymour Benzer, "Monoclonal Antibody Cross-Reactions Between Drosophila and Human Brain," *Proceedings of the U.S. National Academy of Sciences* 80 (1983): 7641–5.

184 **"Herbes gladly cure":** Quoted in Barkan, *Nature's Work of Art*, 1.

185 **"that we were looking":** Cohen, "The Genomics Gamble," 769.

CHAPTER FIFTEEN: THE LORD'S MASTERPIECE

187 **"The Gods are here":** Heraclitus, quoted in M. Delbrück, "Aristotle-totle-totle," in J. Monod and E. Borek, eds. *Of Microbes and Life* (New York: Columbia University Press, 1971), 52.

**Kyriacou knew that repetitions:** My description of this work is based on interviews with Kyriacou and colleagues, and on their papers, including: M. A. Castiglione-Morelli et al., "Conformational Study of the Thr-Gly Repeat in the *Drosophila* Clock Protein, PERIOD," *Proceedings of the Royal Society of London B* 260 (1995): 155–63; C. P. Kyriacou et al., "Evolution and Population Biology of the *period* Gene," *Seminars in Cell and Developmental Biology* 7 (1996): 803–10; and Lesley Sawyer et al., "Natural Variation in a *Drosophila* Clock Gene and Temperature Compensation," *Science* 278 (1997): 2117–20.

189 **"Attempting to study":** Lawrence, *Making of a Fly*, 180.

**"A defense from heat":** Ecclesiasticus 34: 10.

**Blind Watchmaker:** Richard Dawkins, *The Blind Watchmaker* (New York: W. W. Norton & Company, 1987).

190 **evolved in a hot climate:** For a popular article about human sleep and our need for more of it, see Verlyn Klinkenborn, "Awakening to Sleep," *New York Times Magazine* (January 5, 1997): 26.

**Siwicki harvested:** K. Siwicki et al., "Antibodies to the *period* Gene Product of *Drosophila* Reveal Diverse Tissue Distribution and Rhythmic Changes in the Visual System," *Neuron* 1 (1988): 141–50.

191 **Hardin:** Paul E. Hardin, Jeffrey C. Hall, and Michael Rosbash, "Feedback of the *Drosophila period* Gene Product on Circadian Cycling of Its Messenger RNA Levels," *Nature* 343 (1990): 536–40.

**found a new mutant:** Amita Sehgal et al., "Loss of Circadian Behavioral Rhythms and *per* RNA Oscillations in the *Drosophila* Mutant *timeless*," *Science* 263 (1994): 1603–6; Leslie B. Vosshall et al., "Block in Nuclear Localization of *period* Protein by a Second Clock Mutation, *timeless*," *Science* 263 (1994): 1606–9.

**managed to clone the gene:** Michael P. Myers et al., "Positional Cloning and Sequence Analysis of the *Drosophila* Clock Gene, *timeless*," *Science* 270 (1995): 805–8.

191 **found a mouse with something wrong**: M. H. Vitaterna et al., "Mutagenesis and Mapping of a Mouse Gene, *Clock,* Essential for Circadian Behavior," *Science* 264 (1994): 719–25.

**cloned *Clock***: Marina P. Antoch et al., "Functional Identification of the Mouse Circadian *Clock* Gene by Transgenic BAC Rescue," *Cell* 89 (1997): 655–67.

192 ***dClock***: For reviews of the explosion of research and synthesis that followed, see Ueli Schibler, "New Cogwheels in the Clockworks," *Nature* 393 (1998): 620–1; and Steven M. Reppert, "A Clockwork Explosion!," *Neuron* 21 (1998): 1–4.

193 **a molecular biologist in Switzerland**: Schibler, "New Cogwheels."

194 **many more that intermesh**: Jeffrey M. Friedman, "The Alphabet of Weight Control," *Nature* 385 (1997): 119–20.

**A normal *Huntington* gene**: Xiao-Jiang Li et al., "A *Huntington*-associated Protein Enriched in Brain with Implications for Pathology," *Nature* 378 (1995): 398–402; Yvon Trottier et al., "Polyglutamine Expansion as a Pathological Epitope in Huntington's Disease and Four Dominant Cerebellar Ataxias," *Nature* 378 (1995): 403–6.

195 **"I mean it to stand"**: Anthony Burgess, *A Clockwork Orange* (New York: W. W. Norton & Company, 1987 [1962]), preface.

**"The question is"**: Ibid., 83.

196 **"What does God want?"**: Ibid., 95.

197 ***fruitless* was mapped**: Donald A. Gailey and Jeffrey C. Hall, "Behavior and Cytogenetics of *fruitless* in *Drosophila melanogaster:* Different Courtship Defects Caused by Separate, Closely Linked Lesions," *Genetics* 121 (1989): 773–85.

**great generality**: Lisa C. Ryner et al., "Control of Male Sexual Behavior and Sexual Orientation in *Drosophila* by the *fruitless* Gene," *Cell* 87 (1996): 1079–89. For reviews of the story as it unfolded, see Jean Marx, "Tracing How the Sexes Develop," *Science* 269 (1995): 1822–24; and Paul Burgoyne, "Fruit(less) Flies Provide a Clue," *Nature* 381 (1996): 740–1.

198 **"Behold, we put"**: Epistle of James, 3.

**Huxley warned**: Julian Huxley, foreword, *King Solomon's Ring,* ix.

CHAPTER SIXTEEN: PAVLOV'S HAT

200 **"If knowledge isn't self-knowledge"**: Tom Stoppard, *Arcadia* (London: Faber and Faber, 1993), 61.

**found more slow learners**: For a review, see E. O. Aceves-Piña et al., "Learning and Memory in *Drosophila,* Studied with Mutants," *Cold Spring Harbor Symposium in Quantitative Biology* 48 (1983): 831–40.

**"essentially the same"**: R. J. Greenspan, "Flies, Genes, Learning and Memory," *Neuron* 15 (1995): 747.

201 **"we will need"**: Ronald Booker and William G. Quinn, "Conditioning of Leg Position in Normal and Mutant *Drosophila,*" *Proceedings of the U.S. National Academy of Sciences* 78, no. 6 (1981): 3940–4.

**When Tully was a boy**: Tully tells the story in John B. Connolly and Tim Tully, "You Must Remember This," *The Sciences* (May–June 1996): 37–42.

202 **discern varieties of stupidity:** Tim Tully and William G. Quinn, "Classical Conditioning and Retention in Normal and Mutant *Drosophila melanogaster*," *Journal of Comparative Physiology* 157 (1985): 263–77.

**stuck a microelectrode:** For a review, see Eric Kandel, "Nerve Cells and Behavior," *Scientific American* 223 (1970): 57–70.

203 **molecules that compose the message:** Reviewed in E. R. Kandel, "Small Systems of Neurons," *Scientific American* 241 (1979): 67–76.

204 **remarkable genetic engineering projects:** For background, see T. Tully et al., "Genetic Dissection of Consolidated Memory in *Drosophila*," *Cell* 79 (1994): 35–47. For a review, see David A. Frank and Michael E. Greenberg, "CREB: A Mediator of Long-Term Memory from Mollusks to Mammals," *Cell* 79 (1994): 5–8. The experiment itself is reported in two papers: J. C. P. Yin et al., "Induction of a Dominant Negative CREB Transgene Specifically Blocks Long-Term Memory in *Drosophila*," *Cell* 79 (1994): 49–58; and J. C. P. Yin et al., "CREB as a Memory Modulator: Induced Expression of a dCREB2 Activator Isoform Enhances Long-Term Memory in *Drosophila*," *Cell* 81 (1995): 107–15. As Greenspan writes in "Flies, Genes, Learning and Memory," "The rest is history."

"**For flies in the wild**": Ibid., 747.

206 **The mouse genome:** Lee M. Silver, *Mouse Genetics: Concepts and Applications* (New York: Oxford University Press, 1995).

**Silva tested a strain:** The mice were deficient in long-term memory when they were deficient for CREB; see Roussoudan Bourtchuladze et al., "Deficient Long-Term Memory in Mice with a Targeted Mutation of the cAMP-responsive Element-binding Protein," *Cell* 79 (1994): 59–68. Note that Silva's group was able to engineer a mouse with an extra-bad memory but not a mouse with an extra-good memory.

**Kandel and his group:** Cristina M. Alberini et al., "A Molecular Switch for the Consolidation of Long-Term Memory: cAMP-inducible Gene Expression," *Annals of the New York Academy of Sciences* 758 (1995): 261–86.

207 "**We have nothing against universities**": James Barron, "Letters from Serial Bomber Sent Before Blast," *New York Times* (April 26, 1995): A1.

209 **Just as Pavlov's work:** Pavlovian conditioning remains Russia's only serious contribution to the study of behavior, partly because "Mendelo-Morganism" was crushed there early on; see James L. Gould, "Review of Russian Contributions to Invertebrate Behavior, Edited by Charles I. Abramson, Zhanna P. Shuranova, and Yuri M. Burmistrov," *American Scientist* 85 (November–December 1997): 572–4.

CHAPTER SEVENTEEN: ROUGH MOUNTAIN

212 "**Felicity is**": Thomas Hobbes, *Leviathan*, 1651; quoted in David Denby, *Great Books* (New York: Simon & Schuster, 1996), 208.

216 "**Darwin's theory**": Quoted in Walter Gratzer, "Per Ardua ad Stockholm: Review of *I Wish I'd Made You Angry Earlier: Essays on Science, Scientists, and Humanity*, by Max Perutz," *Nature* 393 (1998): 640–1.

217 "**Marxism may be discredited**": Ibid.

217 **Lewontin's views:** See R. C. Lewontin, Steven Rose, and Leon J. Kamin, *Not in Our Genes: Biology, Ideology, and Human Nature* (New York: Pantheon Books, 1984); R. C. Lewontin, "The Dream of the Human Genome," *New York Review of Books* 39, no. 10 (1992): 31–40; and R. Lewontin, *Human Diversity* (New York: Scientific American Books, 1982).

**"filled with the violence":** R. C. Lewontin, "Women Versus the Biologists: Review of *Exploding the Gene Myth,* by Ruth Hubbard and Elijah Wald, and Other Books," *New York Review of Books* (April 7, 1994): 31–5.

220 **"a magnificent structure":** Wilson quotes this passage from Francis Bacon in *Consilience: The Unity of Knowledge* (New York: Alfred A. Knopf, 1998), 23.

221 **"a more subdued draft":** S. Benzer to M. Delbrück, February 3, 1955, Caltech Archives.

CHAPTER EIGHTEEN: THE KNOT OF OUR CONDITION

223 **"The knot of our condition":** Pascal, *Pensées,* 36.

**had to be retracted:** For a critical survey of the battlefield, see John Horgan, "Eugenics Revisited," *Scientific American* (June 1993): 123–31.

225 **Dean Hamer:** This summary of Hamer's work is based on interviews with Hamer and on his own reports. See Dean H. Hamer et al., "A Linkage Between DNA Markers on the X Chromosome and Male Sexual Orientation," *Science* 261 (1993): 321–7; Stella Hu et al., "Linkage Between Sexual Orientation and Chromosome Xq28 in Males but Not in Females," *Nature Genetics* 11 (1995): 248–6. For a popular account, see Dean Hamer and Peter Copeland, *The Science of Desire* (New York: Simon and Schuster, 1994). See also their second book, *Living with Our Genes* (New York: Doubleday, 1998).

226 **a social invention:** Jonathan N. Katz, *The Invention of Heterosexuality* (New York, Dutton, 1995).

**"Drive out nature":** Horace, Epistles, I, x, 24; quoted by Voltaire in the entry on "Character" in his *Philosophical Dictionary,* vol. 1, Peter Gay, trans. (New York: Basic Books, 1962), 125.

**"we smoothe down":** Ibid.

**Twin studies:** J. M. Bailey and R. C. Pillard, "A Genetic Study of Male Sexual Orientation," *Archives of General Psychiatry* 48 (1991): 1089–96.

**brains of gay and straight men:** Simon LeVay, *The Sexual Brain* (Cambridge, Mass.: MIT Press, 1993); S. LeVay and D. H. Hamer, "Evidence for a Biological Influence in Male Homosexuality," *Scientific American* 270 (1994): 44–9.

227 **"Xq28—Thanks for the genes":** Hamer, *Science of Desire,* 21.

**a serious charge:** Eliot Marshall, "NIH's 'Gay Gene' Study Questioned," *Science* 268 (1995): 1841.

**He was cleared:** "No Misconduct in 'Gay Gene' Study," *Science* 275 (1997): 1251.

228 **a gene called *pollux*:** Shang-Ding Zhang and Ward F. Odenwald, "Misexpression of the *white* (*w*) Gene Triggers Male-Male Courtship," *Proceedings of the U.S. National Academy of Sciences* 92 (1995): 5525–9.

with the headline wreathed in a circle: Larry Thompson, "Search for a Gay Gene," *Time* (June 12, 1995): 60–1.

"It's completely silly": John Travis, "Bisexual Bugs," *Science News* 148 (1995): 13–14.

230 a few of the stranger names: Nicholas Wade, "Now Playing at a Nearby Lab: 'Revenge of the Fly People,' " *New York Times* (May 20, 1997): C1.

231 a call from Israel: The report that resulted from that call is Jonathan Benjamin et al., "Population and Familial Association Between the D4 Dopamine Receptor Gene and Measures of Novelty Seeking," *Nature Genetics* 12 (1996): 81–4. See also Hamer, "Thrills," in *Living with Our Genes*, 27–54.

232 Their behavior can be predicted: Jonathan Flint et al., "A Simple Genetic Basis for a Complex Psychological Trait in Laboratory Mice," *Science* 269 (1995): 1432–5.

233 "Maybe it is appropriate": Natalie Angier, "Variant Gene Tied to a Love of New Thrills," *New York Times* (January 2, 1996): A1.

the pursuit of happiness: Klaus-Peter Lesch et al., "Association of Anxiety-related Traits with a Polymorphism in the Serotonin Transporter Gene Regulatory Region," *Science* 274 (1996): 1527–31. Hamer wrote a commentary on that paper: "The Heritability of Happiness," *Nature Genetics* 14 (1996): 125–6. See also Hamer, "Worry," in *Living with Our Genes,* 55–86.

234 "Everybody will be happier": Quoted in Faye Flam, "Pursuing Key to Happiness, Researchers Look for Genes," *Philadelphia Inquirer* (October 4, 1996): A1.

She nurses them: Jon Cohen, "Does Nature Drive Nurture?" *Science* 273 (1996): 577–8. For a related story, see Robert M. Sapolsky, "The Importance of a Well-groomed Child," *Science* 277 (1997): 1620–1.

its whiskers are untrimmed: Nicholas Wade, "First Gene for Social Behavior Identified in Whiskery Mice," *New York Times* (September 9, 1997): C4.

One maggot, when it crawls: J. Steven de Belle, Arthur J. Hilliker, and Marla B. Sokolowski, "Genetic Localization of *foraging* (*for*): A Major Gene for Larval Behavior in *Drosophila melanogaster,*" *Genetics* 123 (1989): 157–63.

236 A computer operator: Maurice Leroy, "Frenchman Is New Heir in Ethiopia," *Philadelphia Inquirer* (November 29, 1996): A12.

239 Today students of Silver's: See J. L. Pierce et al., "A Major Influence of Sex-Specific Loci on Alcohol Preference in C57BL/6 and DBA/2 Inbred Mice," *Mammalian Genome* 9 (1998): 942–48.

CHAPTER NINETEEN: PICKETT'S CHARGE

240 "Human knowledge will be": J. Henri Fabre, *The Insect World of J. Henri Fabre,* John Elder, ed., Alexander Teixeira de Mattos, trans. (Boston: Beacon Press, 1991), 326.

a fine fall afternoon: Saturday, October 29, 1994.

241 dawn call from Stockholm: Wieschaus shared the 1995 prize with Chris-

tiane Nüsslein-Volhard and Edward Lewis for their work on genes and development in fruit flies.

241 **"Molecular biology has no history"**: Benno Müller-Hill, quoted in Sydney Brenner, "A Night at the Operon. Review of *The lac Operon: A Short History of a Genetic Paradigm,* by Benno Müller-Hill," *Nature* 386 (1997): 235.

242 **"I hold the somewhat"**: Ibid.

**"Much of what passes"**: Wilson, *Consilience,* 182–3.

**"laid down"**: Shine and Wrobel, *Morgan,* 2.

**use an electronic device**: Tom Strachan et al., "A New Dimension for the Human Genome Project: Towards Comprehensive Expression Maps," *Nature Genetics* 16 (1997): 126–32.

244 **"the climax of the climax"**: George R. Stewart, *Pickett's Charge: A Microhistory of the Final Attack at Gettysburg, July 3, 1863* (Boston: Houghton Mifflin Company, 1959), ix.

**looking into the far future**: Lee M. Silver, *Remaking Eden: Cloning and Beyond in a Brave New World* (New York: Avon Books, 1997).

**"We have reached"**: Wilson, *Consilience,* 277.

246 **"Nature uses only"**: Quoted in Gleick, *Genius,* 13.

248 **visual awareness**: Francis Crick, *The Astonishing Hypothesis: The Scientific Search for the Soul* (New York: Charles Scribner's Sons, 1994); F. Crick, "Visual Perception: Rivalry and Consciousness," *Nature* 379 (1996): 485.

249 **"If you will be"**: Ibid.

250 **"no longer Gage"**: Richard M. Restak, *The Brain* (New York: Bantam Books, 1984), 147–9.

**blindsight**: Jon H. Kaas, "Vision Without Awareness," *Nature* 373 (1995): 195; Alan Cowey, "Blindsight in Real Sight," *Nature* 377 (1995): 290–1.

251 **"if the atoms never swerve"**: Lucretius, *On the Nature of the Universe,* book 2, lines 250–255, R. E. Latham, trans. (Penguin Books, 1994), 44.

253 **"by incontrovertible direct experience"**: Schrödinger, *What Is Life?,* 92.

**"and that what seems"**: Ibid., 95. Compare Daniel Dennett, *Consciousness Explained* (Boston: Little, Brown and Company, 1991.)

**"Nor will there ever be"**: Schrödinger, *What Is Life?,* 96. Benzer marked this line when he read *What Is Life?* as a young physicist.

**"A human behavior pattern"**: Abraham J. Heschel, *Who Is Man?* (Stanford, Calif.: Stanford University Press, 1965), 9–10.

**"Diversity is as wide"**: Pascal, *Pensées,* 193.

254 **"Thoughts come at random"**: Ibid., 190.

**"Thoughts come into"**: Emerson, "The Over-Soul," in *Essays and Lectures,* 395.

**"I am lying in bed"**: William James, *Talks to Teachers,* Chapter 15, "The Will," in Bruce Kuklick, ed., *Writings 1878–1899* (New York: The Library of America, 1987), 810–11.

**"The grand edifice of Science"**: Max Delbrück, "Homo Scientificus According to Beckett," in W. Beranek, ed, *Science, Scientists, and Society* (New York: Bogden and Quigley, 1972).

255 **And in the tree of life**: This paragraph grew out of a conversation I had

with the writer Lawrence Weschler, who mused about the fractal similarity of the branches.

**"The things we thought"**: Euripides, "The Bacchae," in *The Bacchae and Other Plays*, Philip Vellacott, trans. (London: Penguin Books, 1954), 228.

256 **"Unless one understands the elements"**: Konrad Lorenz, *On Aggression*, Marjorie Kerr Wilson, trans. (New York: Bantam Books, 1966), xi.

**scribbled in his draft**: R. J. Greenspan, J. C. Hall, and T. Tully, *Genes and Behavior: A New Synthesis*, draft of a book for Princeton University Press, 1995.

**"the great parliament of instincts"**: Lorenz, *On Aggression*, 81.

257 **"What is your aim in philosophy?"**: Ludwig Wittgenstein, *Philosophical Investigations*, 3d ed., G. E. M. Anscombe, trans. (New York: The Macmillan Company, 1958 [1953]), 309.

**"like crack through cup"**: Rainer Maria Rilke, "The Eighth Elegy," in *Duino Elegies*, J. B. Leishman and Stephen Spender, trans. (New York: W. W. Norton & Company, 1963), 71.

**"Who's turned us around"**: Ibid.

**the same with consciousness**: Several philosophers see hints in that direction. See, e.g., Daniel C. Dennett, "Our Mind's Chief Asset: Review of *Being There: Putting Brain, Body, and World Together Again*," *Times Literary Supplement* (May 16, 1997): 5.

**"Denmark's a prison"**: W. Shakespeare, *Hamlet*, Act II, Scene 2, in *Comedies and Tragedies*, vol. 2, 617.

# Acknowledgments

In 1991, a team of biologists took a gene from one species and transferred it to a second. When I read the report in the journal *Science,* the experiment interested me because it was quirky: the animals were flies, and the gene changed the flies' sense of time. The experiment also interested me because it opened possibilities that I had assumed—that most of the world had assumed—still lay far ahead in the third millennium.

In the spring of 1994, I had lunch with Ralph J. Greenspan, who is now at the Neurosciences Institute in San Diego. Talking with Greenspan led me to his former teacher Jeff Hall of Brandeis, one of the authors of the report. Talking with Hall led me to his former teacher Seymour Benzer, whose experiments in a Fly Room at Caltech had opened what Benzer called the genetic dissection of behavior. By then I understood that the study of genes and behavior was further along than I had imagined and that the approach Benzer had started—genetic dissection, starting with single genes and working up to behavior—was a tradition of science that would soon be better known. Other studies of behavior worked from the outside in; Benzer and his students worked from the inside out. They had been doing this work ever since the 1960s, and yet until I read that report in *Science* I had been unaware of the genetic dissection of behavior.

At first, I planned to write a book centered on that single experiment. My working title was *A Sense of Time.* But the more I learned, the more I realized that the story really began at the turn of the twentieth century. The school that Benzer started is part of a tradition that has been separate from psychology, ethology, and other schools of behavior in the twentieth century, and is likely to transform the study of behavior in the twenty-first. So the scope of the book widened and the research lengthened until my working title began to take on some uncomfortable ironies.

Ralph Greenspan, Jeff Hall, and Seymour Benzer were extremely generous with their time and hospitality. It is a pleasure to be able to thank each of them here for all the help they gave me while I was writing this book.

After my first year of research, Richard Preston put me in touch with people at the Alfred P. Sloan Foundation. I could not have finished the book without their support. Many thanks to Doron Weber of the foundation for his interest and encouragement.

At Princeton University, Arnie Levine welcomed me into the Department of

Molecular Biology as a Visiting Fellow in 1995, and, as my project lengthened, he allowed me to stay on through 1997. Friends there helped me observe one of the world's leading centers for the study of molecular biology. They put me in the center of things and arranged for me to meet over lunch and dinner with other molecular biologists who were passing through. Special thanks to Alice Lustig, Charles Miller, Tom Shenk, Tom Silhavy, Lee Silver, Shirley Tilghman, Tom Vogt, Evelyn Witkin, and the president of the university, Harold Shapiro. Thanks also to the undergraduate students in Lee Silver's lab, Class of '95.

Shirley Tilghman invited me to teach a writing seminar at Princeton in the spring of 1998, while I was finishing this book. The course allowed me to put together some ideas about writing and science that I had been thinking about for a long time and had been working through, indirectly, with this story. Thanks to Carol Rigolot of the Council for the Humanities for making the course happen.

First Louise Schaeffer and then Nancy Van Doren at Princeton's Biology Library forgave my erratic sense of time. Judith Goodstein of Caltech's Institute Archives was also helpful. Jeff Cramer of the Boston Public Library went beyond the call of duty.

Besides my dozens of hours of interviews with Benzer, I drew on a long unpublished interview that had been conducted by Heidi Aspaturian of Caltech. That oral history was invaluable to me in writing this book. I also drew on unpublished interviews conducted by Garland Allen with Alfred Sturtevant, Carolyn Harding with Max Delbrück, and Horace Freeland Judson with Benzer. I thank them for allowing me to quote from portions of those interviews. Yoshiki Hotta was kind enough to record and send me an oral history of his time in the Benzer lab.

In all, I interviewed nearly 150 biologists, too many to thank here by name. Besides Benzer, Greenspan, and Hall, those who spoke with me at length over many interviews include Steven Helfand, Charabambos Kyriacou, Michael Rosbash, Lee Silver, Tim Tully, and Michael Young. A number of people were particularly helpful in laying out the terrain, including Michael Ashburner, Howard Berg, Sydney Brenner, Francis Crick, Martin Heisenberg, Eric Kandel, Ed Lewis, James Watson, Eric Wieschaus, and E. O. Wilson. Also helpful were historians of biology, including Garland Allen, Angela Creager, Daniel J. Kevles, and Jane Maeinschein. Robert Kohler's history of early fly work, *Lords of the Fly*, was good reading, and gave me ideas for a few of my book's illustrations.

For many different kinds of help, thanks to Neil Beach, Barbie Benzer, Anthony Bonner, Manny Delbrück, Karen Fahrner, David Fleischer, Bob Freidin, Burt Hall, Sue Judd, Carolynne Lewis-Arevalo, Monika Magee, James McPherson, Rosie Mestel, Beth Panzer, Rabbi Sandy Parian, Pam Polloni, Richard Rhodes, James Shreeve, Barbara Smith, Norma Deupree Sperry, Rabbi Shira Stern, and the staff of Paganini Ristorante and Cafe. Special thanks to Kathy Robbins.

Benzer volunteered to read the manuscript. So did Reb Brooks, Ralph Greenspan, Jeff Hall, Don Herzog, David and Mair La Touche, Laurie Miller, Chip Quinn, Michael Rosbash, Keith Sandberg, Lee Silver, Shirley Tilghman, Tim Tully, and Mike Young. Some of them read more than one draft, and they made many useful comments, suggestions, and corrections. Greenspan pulled an all-nighter. Benzer and Hall read the galleys. John Tyler Bonner read every draft, always with that combination of enthusiasm and levelheadedness that is cherished by his worldwide circle of friends.

Many thanks once again to my agent, Victoria Pryor, and my editor, Jon Segal, who had more patience and more faith in me and in my project than I did myself. Thanks to Ida Giragossian and Michael Rockcliff at Knopf.

I thank my friends and family for helping me clarify my thoughts about genes and behavior and for reminding me that there are other ways of looking at life too. Aaron and Benjamin talked over the project during family dinners and waited patiently when it got in the way of family adventures.

My wife, Deborah Heiligman, was more enthusiastic about this one than about any of my other books. During the last stretch, when I often worked through the night, she slept in my study to keep me company, though her own writing sometimes suffered in the morning. This book would not have been what it is without her help.

Time, love, memory, as seen through a compound eye.

# Index

# Illustration Credits

14  After Niko Tinbergen, "Social Releasers and the Experimental Method Required for Their Study," *Wilson Bulletin* 60 (1948): 6–52.
25  Courtesy of the American Philosophical Society Library.
29  From T. H. Morgan, "Localization of the Hereditary Material in the Germ Cells," *Proceedings of the U.S. National Academy of Sciences* I (1915): 420–29.
30  Courtesy of the American Philosophical Society Library.
32  Courtesy of the American Philosophical Society Library.
36  Courtesy of Seymour Benzer.
    Courtesy of Seymour Benzer.
41  Courtesy of Cold Spring Harbor Laboratory Archives.
42  Courtesy of Seymour Benzer.
43  Courtesy of Seymour Benzer.
48  Courtesy of Manny Delbrück.
49  Courtesy of Ross Madden, Black Star, and Cold Spring Harbor Laboratory Archives.
50  Courtesy of Cold Spring Harbor Laboratory Archives.
51  From Seymour Benzer, "Genetic Fine Structure," *Harvey Lectures* 56 (1960): 1–21.
56  Courtesy of Seymour Benzer.
75  Courtesy of the Cajal Institute.
80  Francis Galton, *Inquiries into Human Faculty,* 120.
81  Courtesy of Seymour Benzer.
88  Courtesy of the American Philosophical Society Library.
90  Courtesy of the Syndics of Cambridge University Library.
91  From T. H. Morgan, C. B. Bridges, and A. H. Sturtevant, "The Genetics of *Drosophila,*" *Bibliographia Genetica* 2 (1925): 1–262.
111  Courtesy of California Institute of Technology Archives.
121  T. H. Morgan and C. B. Bridges, "The Origin of Gynandromorphs," in *Contributions to the Genetics of Drosophila melanogaster* (Carnegie Institute of Washington, 1919): 1–122.
123  From Seymour Benzer, "Genetic Dissection of Behavior." Courtesy of *Scientific American* and the estate of Bunji Tagawa.

133   Courtesy of Seymour Benzer.
150   Courtesy of Seymour Benzer.
      Photograph of Hall and Rosbash, courtesy of Karen Fahrner.
      Photograph of Benzer and poster from *The Fly*, courtesy of Jeff Hall.
156   Courtesy of Chris Hadfield.
182   Courtesy of Edward Lewis.
184   Courtesy of Seymour Benzer.
210   Courtesy of Eleana Savvateeva and Tim Tully.
      Courtesy of Barbie Benzer.
214   Courtesy of Seymour Benzer.
      Photograph of Feynman and Benzer, courtesy of California Institute of Technology Archives.
      Photograph of Benzer and Watson, courtesy of Cold Spring Harbor Laboratory Archives.
235   Electron photomicrograph courtesy of Edward Lewis.
247   Courtesy of Laurent Seroude and Seymour Benzer.
249   Courtesy of the New York Academy of Medicine Library.

Jonathan Weiner worked as a writer and editor at *The Sciences*. He is the author of *Planet Earth, The Next One Hundred Years,* and *The Beak of the Finch,* which won both the Los Angeles Times Book Prize and the Pulitzer Prize. During the writing of this book he was Visiting Fellow in the Department of Molecular Biology at Princeton University, and then McGraw Professor in Writing. He lives in Bucks County, Pennsylvania, with his wife and their two sons.

A NOTE ON THE TYPE

This book was set in Fairfield, the first typeface from the hand of the distinguished American artist and engraver Rudolph Ruzicka (1883–1978). In its structure Fairfield displays the sober and sane qualities of the master craftsman whose talent has long been dedicated to clarity. It is this trait that accounts for the trim grace and vigor, the spirited design and sensitive balance, of this original typeface.

Rudolph Ruzicka was born in Bohemia and came to America in 1894. He set up his own shop, devoted to wood engraving and printing, in New York in 1913 after a varied career working as a wood engraver, in photoengraving and banknote printing plants, and as an art director and freelance artist. He designed and illustrated many books, and was the creator of a considerable list of individual prints—wood engravings, line engravings on copper, and aquatints.

*Composed by North Market Street Graphics, Lancaster, Pennsylvania*
*Printed and bound by Quebecor Printing, Martinsburg, West Virginia*
*Designed by Robert C. Olsson*